ENVIRONMENTAL SCIENCES AND APPLICATIONS

Series Editors: ASIT K. BISWAS
MARGARET R. BISWAS

Volume 9

EROSION AND ENVIRONMENT

ENVIRONMENTAL SCIENCES AND APPLICATIONS

Other titles in the series

NOTICE TO READERS

Dear Reader

If your library is not already a standing order customer or subscriber to this series, may we recommend that you place a standing or subscription order to receive immediately upon publication all volumes published in this valuable series. Should you find that these volumes no longer serve your needs your order can be cancelled at any time without notice.

The Editors and the Publisher will be glad to receive suggestions or outlines of suitable titles, reviews or symposia for consideration for rapid publication in this series.

ROBERT MAXWELL
Publisher at Pergamon Press

EROSION AND ENVIRONMENT

MILOŠ HOLÝ

Professor at the Technical University of Prague, Czechoslovakia

Translated by
JANA ONDRÁČKOVÁ

PERGAMON PRESS

OXFORD · NEW YORK · TORONTO · SYDNEY · PARIS · FRANKFURT

U.K.	Pergamon Press Ltd., Headington Hill Hall, Oxford OX3 0BW, England
U.S.A.	Pergamon Press Inc., Maxwell House, Fairview Park, Elmsford, New York 10523, U.S.A.
CANADA	Pergamon of Canada, Suite 104, 150 Consumers Road, Willowdale, Ontario M2J 1P9, Canada
AUSTRALIA	Pergamon Press (Aust.) Pty. Ltd., P.O. Box 544, Potts Point, N.S.W. 2011, Australia
FRANCE	Pergamon Press SARL, 24 rue des Ecoles, 75240 Paris, Cedex 05, France
FEDERAL REPUBLIC OF GERMANY	Pergamon Press GmbH, 6242 Kronberg-Taunus, Hammerweg 6, Federal Republic of Germany

First edition 1980

British Library Cataloguing in Publication Data
Holý, Miloš
Erosion and environment. - (Environmental sciences and applications; vol. 9).
1. Soil erosion
I. Title II. Series
631.4'5   S623   79-41230
ISBN 0-08-024466-1

Printed and bound in Great Britain by
Redwood Burn Limited
Trowbridge & Esher

Contents

Contents

Acknowledgement

The author would like to express his acknowledgement to the translator, Mrs. Jana Ondráčková, for her excellent cooperation.

1. Introduction

The development of society is determined by its capacity to exploit the resources of the biosphere. Some of these resources may in time be exhausted or deteriorated.

The soil as one of the main resources of the biosphere has been defined by the International Soil Science Society as follows: "The soil is a limited and irreplaceable resource and the growing degradation and loss of soil means that the expanding population in many parts of the world is pressing this resource to its limits. In its absence the biospheric environments of man will collapse with devastating results for humanity".

The intensive exploitation of soil for agricultural production and capital construction in almost all branches of the national economies of most countries have gradually disturbed the natural soil cover and have exposed the soil surface to the action of erosion agents, i.e., the destructive effects of water and wind on the soil surface, the dislocation and removal of soil matter from the earth surface and its deposition in areas where the activity of these agents is decreasing (Fig. 1).

The action of water, wind and glaciers which under natural conditions proceeded slowly, and from the point of view of one human generation went unnoticed, have been significantly accelerated by man's activities with consequences that are in many cases highly unfavourable for human society. Historical erosion which in past geological periods partook of the formation of the earth's relief, in our time has changed to current erosion which models significantly the surface of this planet. It appears as normal erosion when the erosion processes take place slowly preserving the balance in natural ecosystems, and as accelerated or abnormal erosion when the balance in natural ecosystems is disturbed or destroyed. Accelerated erosion is the cause of the dangerous dislocation and removal of soil particles and chemical substances.

In normal erosion the loss of soil particles is set off by the formation of new soil particles from the subsoil; the processes of the removal of soil particles are slow and almost unnoticeable. Accelerated erosion causes a restructuring of the earth surface by the wash of soil particles and nutrients which can no longer be resupplied by the soil formation process. In some cases chemical substances supplied to the soil are completely removed. The unfavourable consequences of accelerated erosion which have recently been significantly strengthened by industrialization and urbanization processes do not only pose a threat to the soil but also to water which is another basic natural resource being polluted by substances moved by erosion.

1

Fig. 1.   Destructive effect of erosion on agricultural land
          (photo by courtesy of the Soil Conservation Service,
          USA).

1.1   EFFECTS OF EROSION ON THE NATIONAL ECONOMY

Erosion affects a number of branches of the national economy.

1.1.1 *Consequences of Soil Loss*

Agriculture is that branch of the national economy which is most affected by the
erosion processes. The detachment and transportation of soil particles often
occurs on a large scale. There are frequent cases of the denudation of the subsoil
caused by intensive rainfall which washes away the shallow top soil layer. This
has extremely unfavourable consequences for agriculture and forestry considering
the long-term process of soil formation. The decrease in soil fertility resulting
from the loss of soil particles depends on the type of soil and on the depth of
the soil profile. Surveys carried out in deep soils in the grain growing areas
of the USA[14] have shown that the loss of 50.8 mm of soil reduced fertility by
15%, 101.6 mm by 22%, 152.4 mm by 30%, 203.2 mm by 41%, 254 mm by 57% and 304.8 mm
by 75%.

Soil fertility is reduced by the removal of plant nutrients. It is very difficult to determine the quantitative values of this removal. This is because decreased soil fertility resulting from this removal depends on the amount, type and form of nutrients supplied and on the properties of the respective soil. Investigations carried out in various countries, show that the loss of nutrients from agricultural land is considerable and has become a serious problem for agricultural production. R. P. Beasly[1] estimates that US \$6.8 billion worth of N, P_2O_5 and K_2O are lost in the US every year without any use as a result of erosion processes. M. Holý[7] has found by surveys carried out in the vineyards in North Bohemia that the annual loss of humus (expressed in C) reaches 2.000 kg per ha, the annual loss of P_2O_5 amounts to 30% and the loss of K_2O to 20% of the total amount supplied per year. The loss of plant nutrients not only reduced crop yields but also worsened the quality of the crop.

Lower intensity erosion processes cause the loss of fine soil particles. This changes the soil structure and texture and reduces the water capacity of soils. In high intensity water erosion processes a considerable proportion of the topsoil is washed away and rain water does not as a rule reach the lower horizons of the soil profile with a lower content of organic substances and lower permeability; the soil profile is not adequately supplied with moisture which in dry seasons will have an adverse effect on the development of the vegetation.

The removal of soil particles by wind erosion often denudes the roots of the vegetation which then wilts and dries. This is the consequence of dust storms.

Deep rills and gullies caused by water erosion cut agricultural land into uneven plots reducing the possibility of efficient mechanized tillage.

1.1.2 Consequences of Soil Removal and Sedimentation

Soil particles detached by surface runoff are deposited at the foot of slopes following the decrease of the tangential stress of the water. The fine material is further transported by water into the hydrographic system.

These fine soil particles silt natural and artificial waterways (navigation, irrigation, drainage and other canals) and water reservoirs, and clog structures on water flows. They reduce the required water discharge in water courses and canals, thereby affecting the water supply to industry, agriculture, etc. In some cases they considerably reduce or limit the function of the canal itself. This negative effect is mainly evident in irrigation canals whose silting significantly affects the operation and economy of irrigation systems.

Silt raises the level of river beds and the beds of water reservoirs increasing the danger of the inundation of the adjoining area, the rise in the groundwater level and the consequent waterlogging of the whole area. Silting reduces the function and service life of structures on water courses, namely of off-take structures.

Silt threatens the operation of storage reservoirs reducing their capacity. With regard to the great number of reservoirs being frequently situated in upstream areas losses caused by silting are a world problem. R. P. Beasley[1] states that more than 33% of the capacity of water reservoirs in the midwestern states of the US is lost by silting within 50 years of their construction. With many reservoirs as much as 5% of volume is lost annually by silting (Fig. 2).

Silting is very quickly evident in small water reservoirs built in upstream areas. In many cases such water reservoirs are put out of operation within a few years.

Fish ponds have been seriously damaged by silting and their capacity reduced in many countries.

Fig. 2.   Silting of water reservoir, South Carolina, USA
          (photo by courtesy of the Soil Conservation Service,
          USA).

The silting of water reservoirs seriously affects the water supply. The consequences of the reduced volume of water reservoirs used for power production are a decline in power production and the clogging of the water turbines.

The transportation of silt often threatens water courses and canals. The costs of dredging are immense and raise the costs of navigation. Torrential streams trans- porting gravel into navigable water courses are very dangerous.

In some mountain areas torrential rains put into motion debris flows which threaten structures and towns (in the Alps they are called "mur", in the Caucasian Mountains "syely"). A special retention reservoir had to be built to protect Alma-Ata in Kazakhstan which in 1973 retained 3 million m^3 of gravel. The flow of debris reached a height of 30 m and had a force of 1200 MJ[15] (Figs. 3 and 4).

Fig. 3. Debris flow in the Caucasian Mountains (photo by
         courtesy of the Institute of Irrigation and
         Drainage, Technical University, Prague).

The recreational value of the eroded area, especially of silted water courses and
water reservoirs is considerably reduced and the eroded banks of water flows and
reservoirs become unsuitable for recreation.

The transportation of soil particles by wind erosion has adverse effects on whole
areas. Debris and soil removed by wind erosion are often deposited on vegetation
and they damage buildings, communications, canals, ditches, etc. (Fig. 5).

Soil particles carried by wind pollute the atmosphere causing a health hazard to
people and animals who suffer from diseases of the respiratory tract, and eye
inflammation. The extent of the danger has been proved by data published by
W. S. Chepil and N. P. Woodruff[10] who found up to 310 tons of dust particles in
1 km^3 of air in a dust storm.

1.1.3 Consequences of the Transportation of Chemical Substances

The traditional approach to erosion processes, i.e., the assessment of erosion as
a factor which devastates the soil surface and is the source of silt and debris

clogging water courses and reservoirs, is not broad enough to permit the full evaluation of the problem. The soil comes into contact with vast quantities of
chemicals of various types and levels of toxicity. The transportation of these
substances owing to erosion poses a grave danger to society all the more because
chemical substances are put into motion very easily, and erosion processes are
scattered over large areas which makes it very difficult to draw up and implement
efficient and economical erosion control measures. Chemical substances infiltrate
surface and groundwaters and limit the use of water resources[6].

Fig. 4.   Flow of debris in the Durundji river, USSR (photo
           by courtesy of the Institute of Irrigation and
           Drainage, Technical University, Prague).

The most frequent sources of these chemicals are chemical fertilizers and different
types of pesticides, herbicides and fungicides applied in large quantities in agriculture as well as industrial and agricultural wastes discharged on, or into the
soil. The chemicals are transported from the site of use or deposition by water
or by wind. In an extensive research programme investigating the movement of
pesticides in the atmosphere[6] the concentrations of DDT and its derivatives were
observed in air flows carried by trade winds from Europe and Africa over Barbados.
The great mobility was observed of these highly toxic substances on a world scale.
It has been observed that 1 m^3 of air contains 7.8 10^{-4} g of these substances.
Similar results have been obtained by researchers studying the transportation of
chemicals by water.

Fig. 5. Deposition of Soil by wind erosion, South Moravia,
Czechoslovakia (photo by courtesy of the Research
Institute for Land Reclamation, Prague).

Chemical substances from the soil surface appear in surface, subsurface and groundwaters with increasing frequency and significantly affect water quality. The
increasing application of fertilizers, namely NPK and the subsequent impact of
this application on the quality of water resources on a world scale has induced the
UN bodies to deal with this problem[6]. In many countries the use of NPK fertilizers
has reached such a degree that they gravely affect water quality in water reservoirs.
Examples of this are the Želivka waterstorage reservoir which supplies drinking
water for Prague and many other water reservoirs in various countries, whose operation is unfavourably influenced by erosion.

The nitrogen and phosphorus levels cause the eutrophication of many water reservoirs,
especially ponds and lakes limiting their use for recreation and other purposes
(Fig. 6).

1.2 GEOGRAPHICAL DISTRIBUTION OF EROSION ON A WORLD SCALE

The geographical distribution of erosion is related to the occurrence of two basic
factors — precipitation and wind as well as on the concurrent occurrence of the
other factors, such as morphology, the vegetative cover, soil properties, etc.

Water erosion is caused by precipitation. Areas with a low precipitation usually
have a small surface runoff, because precipitation water infiltrates into the soil
and is consumed by vegetation. Higher annual precipitation, usually more than
1000 mm, results in the growth of dense vegetation which prevents the development

of erosion. Many authors[9] therefore assume that water erosion is most widespread in areas with medium annual precipitation rates, with intensive agriculture which disturbs the soil cover and in areas with high annual precipitation rates where the soil has been stripped of its natural forest cover. N. Hudson[9] constructed the relationship between water erosion and mean annual rainfall (Fig. 7).

Fig. 6. Eutrophication of water reservoir (photo by J. Říha).

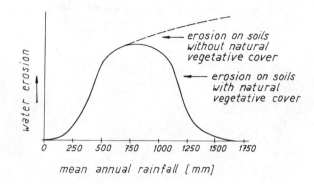

Fig. 7. Dependence of water erosion on mean annual rainfall
         after N. Hudson.

Erosion is affected not only by the amount of rainfall but also by its type. Intensive rainfall is in many cases the major agent of the intensity of erosion
processes. N. Hudson believes that such precipitation occurs mainly between latitudes 40° North and 40° South with exceptions to the rule. The generalized map of the
distribution of water erosion processed from the distribution of rainfall on a world
scale and from data ascertained by the author is shown in Fig. 8[9].

Soil erosion by water is thus most serious between these latitudes with the exception of dry deserts and equatorial forests.

Erosion by wind is caused by the effects of wind on dry soil. N. Hudson[9] believes
that vulnerable regions are those with a low mean annual rainfall, particularly
those with a mean annual rainfall of less than 250-300 mm, and prevailing winds in
one direction associated with large, fairly level land masses lacking any vegetative cover or having insufficient vegetative cover. The largest areas which are
thus threatened by wind erosion are the Great Plains in the USA, the Sahara and
Kalahari deserts in Africa, the central Asia regions (especially the steppe of the
USSR) and central Australia (Fig. 9).

1.2.1 The History of Erosion

The decisive epoch of the development of soil erosion began when man settled down
and began turning pastureland into farmland. The intensive exploitation of the
land disturbed the natural soil vegetative cover and exposed its surface to the
effects of erosive agents. Only rarely did man succeed in overpowering the erosive
agents and to introduce such forms of agriculture that did not destroy the land.
The devastation of land by erosion often led to the downfall of civilizations, e.g.,
in Mesopotamia, Syria, China and elsewhere.

In Mesopotamia, the rivers Euphrates and Tigris were used to irrigate the surrounding
areas. The irrigation system built 2000 years B.C. was the basis for the

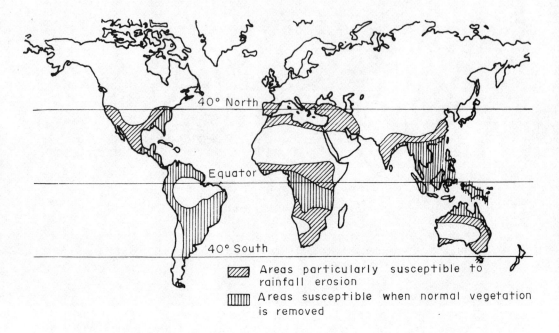

Fig. 8.   Distribution of water erosion on world scale after
          N. Hudson.

establishment of a highly developed civilization. Dense forests covering the
mountain ranges and lining the Euphrates and the Tigris rivers were felled, the
irrigation canals and water courses were silted as a result of erosion and Mesopotamia became a desert. Only the remnants of the once powerful Babylon and its
numerous historical monuments now covered with layers of desert sand have remained
to testify to the existence of the once highly developed country.

Fertile Syria had to sacrifice its dense forests for timber for ship building and
town construction. In the course of several centuries erosion caused immense soil
loss, agriculture was destroyed and sand covered the ruins of the once prosperous
cities. The country which had once supplied Rome with large quantities of wine
and olives has become a desert whose reconstruction requires enormous effort and
investments.

The world famous Lebanon cedars which had once been the pride of that area covering
its mountain ranges were the cause of its devastation. Three thousand years ago
the trees were felled and the timber used for ship building and the construction
of towns and cities and the denuded soil cultivated. Erosion resulted in the devastation of the soil and thereby the devastation of the whole country. Only in
isolated parts of the country where attempts were made to farm on stone terraces
agriculture developed for thousands of years.

The same development may be traced in many other countries.

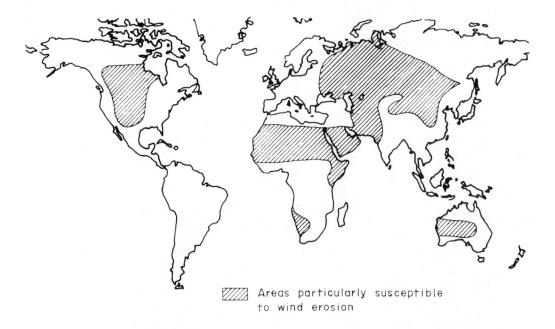

 Areas particularly susceptible
to wind erosion

Fig. 9. Distribution of wind erosion on world scale after
N. Hudson.

In China the territory along the Huang Ho (Yellow River) used to be very fertile
land and was renowned for its riches. Today it is eroded waste land (Fig. 10). The
light colour of the soil washed into the Yellow River has given the river its name.
In the catchment area of the Yellow River 50% of the total weight of the water in
the flood season is silt. In the river delta a network of canals was built in a
total length of more than 600 km. Many of these canals have been built above ground
level and dams have been built to prevent the soil from being washed into the
canals. The dams are often broken through owing to the continuous rise in the
profile level of the flow and have become a grave danger for the farmland in the
surrounding area as well as a threat to the lives of millions of inhabitants in
the area.

The settlement of North America led to serious erosion in the largely untouched
virgin land of the Western Hemisphere. The white settlers who came from western
Europe gradually seized Indian land and introduced alien farming methods which were
totally unsuitable for the new environment. Natural growth in the prairie land
was burnt down and monocultures planted. Farming was gradually mechanized and
intensified which in turn accelerated the process of erosion by water and wind.
Dust storms lasting many days became a common phenomenon. The fine dust particles
prevented the penetration of sun rays, clogged machines and equipments and caused
serious lung diseases. The wind denuded the roots of the vegetation, fine dust
grains were transported hundreds of miles and course soil particles were blown
along the surface as ground drift and often covered houses, farmland, trees, etc.
(Fig. 11). In one such dust storm which hit the Mississippi river valley in 1936[2]
150 tons of soil was blown from the fields mostly into the sea and destroyed more
than 4 million hectares of good farmland in a matter of days. Dust storms often
recurred with varying intensity, threatening especially the states of Colorado,
Kansas, Missouri, Texas and Wyoming.

Fig. 10. Water erosion in China (photo by courtesy of the
          Institute of Irrigation and Drainage, Technical
          University, Prague).

The US Soil Conservation Service survey in 1934[1] showed that there were in the US
more than 20 million hectares of eroded topsoil and that this area was gullied to
such an extent that it was completely useless for agriculture. Some 112 million
hectares of land had been devastated to such an extent that it had become uneconomic
for agriculture and cattle breeding. Significant damage had also been done to
forest soils, mainly in areas with dense housing construction, highways, mining
areas, etc. (Fig. 12). This not only reduces farm production but also threatens
technical structures and is the main source of sediments which pollute and silt
water courses and reservoirs.

The situation is different in countries where ancient cultures led to a close relationship between the population and the land. In the Andes in South America farmers

till the steep slopes of the mountains where their ancestors, the Incas lived and cultivated the land thousands of years ago. The fields were meticulously terraced to intercept rainfall and to prevent the occurrence of erosion processes. This enormous work carried out over such a vast period of time indicates what should be done to protect the soil to feed future generations.

Fig. 11. Dust storm effects in Oklahoma, USA (photo by courtesy of the Institute of Irrigation and Drainage, Technical University, Brno, Czechoslovakia).

In Europe, erosion is evident mainly in the territory of the USSR, affecting cultivated steppe and woodland steppe and the mountainous regions of the Caucasus and the Carpathians[15]. Erosion affects approximately 60 million hectares of farmland in the Ukraine, in Moldavia, in the Russian Federation and in the Volga river basin. Erosion is widespread and very serious in the high altitude Transcaucasian Republics and partly also in Central Asia where the productivity of fields, pastureland and other soils is decreasing.

After S. S. Sobolev[12] in the USSR water erosion causes the annual loss of 535 million tons of soil per year and with it the loss of 1229 thousand tons of N, 539 thousand tons of P_2O_5 and 12135 thousand tons of K_2O.

Wind erosion is very common in these areas. "Black storms" in the Ukraine used to be the cause of starvation. The biggest disaster of this kind hit the Ukraine in

1891. In that year the crop in Russia was so poor that the population of entire villages starved to death. L. N. Tolstoy who witnessed the disaster wrote that in one village only 13 farms remained of the former 70 and that these 13 barely survived. The inhabitants of many villages left their homes and went begging.

Fig. 12. Highway in Colorado, USA, damaged by stream erosion
(photo by courtesy of the Soil Conservation Service,
USA).

The erosion processes in the USSR led the Soviet government to pass a Decision in 1967 on Soil Protection from Wind Erosion. The Decision was immediately put into force and has yielded good results. Under it protective forest belts were planted on 4 million hectares of land in the USSR, in the Ukraine and in Kazakhstan. Further erosion control measures are being implemented.

REFERENCES

1. Beasley, R. P., *Erosion and Sediment Pollution Control*, Iowa, USA, 1972.
2. Bennet, H. H., *Soil Conservation*, McGraw Hill, New York-London, 1939.
3. Bennet, H. H., *Elements of Soil Conservation*, McGraw Hill, New York-London,
    1947.
4. Biswas, A. K., *Water and Environment*, Pergamon Press, 1979.
5. Bučko, Š., Holý, M. and Stehlík, O., Soil Erosion in Czechoslovakia, *Journal
    of the Czechoslovak Geographical Society*, 1964.
6. Economic Commission for Europe, *Proceedings, Seminar on the Pollution of Waters
    by Agriculture and Forestry*, Vienna, 1973.
7. Holý, M., Problems of the Evaluation of Water Erosion Effects, *Rostlinná výroba*
    No.8, 1964.
8. Holý, M., Říha, J. and Sládek, J., *Society and the Human Environment*, Svoboda,
    Praha, 1975.
9. Hudson, N., *Soil Conservation*, Bt Batsford Ltd., London, 1971.
10. Chepil, W. S. and Woodruff, N. P., Sedimentary Characteristics of Dust Storms,
    Am. Journ. Sci., Vol.255, 1957.
11. Pasák, V., Wind Erosion on Soil, *Scientific Monographs*, VÚM, Zbraslav 3, 1970.
12. Sobolev, S. S., *Razvitiye erozionnykh processov na teritorii evropeyskoy chasti
    SSSR i borba s nimi*, Moskva, 1948.
13. Sobolev, S. S., Eroziya pochv v SSSR i borba s nieyu, *Collection Eroziya pochv
    i borba s nieyu*, Moskva, 1957.
14. Stallings, J. H., *Soil Conservation*, Englewood Cliffs, N. J., Prentice-Hall,
    USA, 1964.
15. Štěpa, B., *Land Improvement in the USSR*, Moskva, 1975.
16. Švehlík, R., Dust Storms below the White Carpathians, *Ochrana půdy* No.8, 1968.
17. Zachar, D., *Soil Erosion*, SAV, Bratislava, 1970.
18. Záruba, Q. and Mencl, V., *Engineering Geology*, Academia, Praha, 1974.

2. Classification of Erosion

Erosion (from the Latin *erodere*) is manifested by the deterioration of soil surface effected by exogenous forces, especially water, ice, wind and man as the significant anthropogenic factor. The disturbance of the soil surface is accompanied by the removal of the detached soil particles by the force of kinetic energy of some of the erosion agents, namely water and wind and the deposition of this matter with a decrease in this energy.

2.1 CLASSIFICATION OF EROSION BY EROSION AGENTS

By the agents causing the occurrence and affecting the course of the erosion process, erosion may be classified into:

- erosion by water
- erosion by glacier
- erosion by snow
- erosion by wind
- anthropogenic erosion.

The said types of erosion may occur separately or in combination causing erosion of varying intensity. On the world scale greatest damage to the economy comes from water and wind erosion whose unfavourable effects are increased by anthropogenic erosion.

Water erosion is caused by the kinetic energy of raindrops impinging on the soil surface and by the mechanical force of surface runoff. Surface runoff is caused by heavy rainfall, rainfall of long duration, snow water from spring thaw and the concentration of water in the natural or artificial hydrographic network. Bank erosion is effected by the waters of seas, lakes and ponds (Figs. 13 and 14), groundwaters, namely water from karst formations cause not only mechanical but also chemical erosion. Other types of mechanical erosion are evorsion and abrasion of the bank and bedrock water flows, lakes and seas.

Glacier erosion is caused by a glacier moving by force of gravity into the valleys, thereby eroding the bedrock which it grinds and wears, and cuts by boulders frozen in the glacier. It removes and transports into the valleys detached weathered rocks which when deposited form moraines. The earth and stones deposited in moraines are transported into water flows by water from thawing glaciers considerably contributing to the silting of these flows.

Fig. 13.   Bank erosion — Black Sea in Bulgaria (photo
           J. Říha).

Fig. 14. Bank erosion — storage water reservoir near Brno
          in Czechoslovakia (photo by courtesy of the Water
          Management Research Institute, Brno, Czechoslovakia).

Glacier erosion is restricted to high mountains (the Alps, Caucasian Mountains, the
Rocky Mountains, etc.).

Snow (nival) erosion is caused by the movement of snow in the form of avalanches
whose erosion activity takes place at high pressure and velocity. The avalanche
usually devastates the affected area. Snow erosion may also be caused by the slow
movement of a layer of snow along an unfrozen soil surface during the spring thaw.
This type of erosion mainly affects submontane areas.

Wind erosion is evident by the removal of soil particles by the force of the kinetic
energy of the wind. These soil particles are transported and deposited when the
energy of the wind flow drops.

Data on the effects of different types of erosion show that on a world scale wind
erosion is not such a serious problem as water erosion. There are, however, large
areas where wind erosion causes damage on an equal or even larger scale. Wind erosion is a typical phenomenon in arid and semi-arid regions, its manifestations may
occur in humid areas, namely on drier soils with no vegetative cover or having
unfavourable physical properties.

Man by his activities often sets off the process of erosion and has great influence on its development. He is an important factor in accelerated erosion, be it directly or indirectly. His negative effect is felt by the destruction of the natural vegetative cover of the soil and by replacing the natural vegetation by one that gives the soil small protection, i.e., farm crops, by deteriorating the physical, chemical and biological properties of the soil, by concentrating surface runoff resulting from various agricultural techniques, causing soil pollution by waste disposal, and by the building of technical structures and urbanization.

Intensified agriculture, the building of communications and urbanization rank among the most important causes of anthropogenic erosion.

The intensification of farming leads to the establishment of large monocultures very often planted without any regard to the configuration of the terrain. Planted with vegetation which gives the soil small protection these areas are exposed to the kinetic energy of raindrops and affected by surface runoff which result in intensive erosion, namely on long slopes (Fig. 15). Heavy machinery used on the farms disturbs the soil structure and diminishes the infiltration capacity of the soil. Unsuitably designed road networks form runoff paths for water and are the basis for future rills and gullies. Unsuitably built drainage systems in hilly areas have similar negative effects. Mineral fertilizers, pesticides and other chemicals which are intensively applied to the soil are washed and blown into water resources which they pollute.

Erosion often occurs on irrigated soils, especially due to irrigation return flow from surface irrigation systems. Using surface irrigation methods it is very difficult to eliminate erosion altogether[4]; by sprinkler irrigation it may be done by selecting the irrigation rate according to the infiltration capacity of irrigated soils.

Water resources may be polluted by irrigation return flow, which as a result of erosion has been enriched with various chemical substances from mineral fertilizers, pesticides, salts, etc. In many countries in arid zones return flow from irrigated areas causes soil salination. In the western parts of the USA salination caused by this type of erosion is threatening 18 million hectares of irrigated farmland[3].

In mountainous areas erosion may be caused by overgrazing which results in the denudation of the soil surface. Grazing livestock, especially sheep, trail the land and these trails become runoff paths and rills, which further erosion turns into gullies and ravines (Figs. 16 and 17).

Farmed land must be made accessible by a network of paths and roads allowing the access of heavy machinery. On sloping land such roads are often very steep and water running down these steep roads gains considerable velocity and tangential stress which especially in less resistant soils create deep sunken roads which in time often become gullies and ravines.

Highways, roads and railways also contribute to the development of erosion. Their unconsolidated and often steep embankments and shoulders are significantly disturbed by runoff (Fig. 18). The detached particles are transported into drainage ditches where they are deposited thereby silting and clogging them. Observations made in the USA[10] have shown that heavy rains washed away ten times more soil from the unconsolidated embankments of highways and roads than from cultivated land. This danger should mainly be taken into consideration in the construction of highways.

Urbanization and the construction of housing estates often result in a more intensive erosion than does intensive agriculture. The soil surface is denuded on large areas, it is trampled and compacted to such an extent that it no longer allows water

to infiltrate into the deeper layers, heavy machinery disturbs the surface even
more and grooves the whole building site which results in even more intensive erosion.

Fig. 15. Farmland depleted by water erosion (photo by
courtesy of the Soil Conservation Service, USA).

In Japan agricultural land is extremely valuable and costly housing estates have
been built on infertile mountain slopes. This has caused intensive erosion, the
removal of large amounts of silt and frequent landslides. The construction of the
Johns Hopkins University in the USA[3] removed more than 500 tons ha^{-1} of sediments
per year owing to intensive erosion. With regard to intensively advancing urbanization on a world scale this type of anthropogenic erosion is becoming a very
serious problem indeed.

The above examples show the growing danger and significance of anthropogenic erosion. New types of erosion are appearing and their intensity is increasing, e.g.,
erosion caused by mining (water and wind erosion in areas with open cast mining),
erosion caused by waste disposal heaps, forest passes cut for the construction of

cable-lines and ski-tows in mountain areas, etc. The danger of such erosion and its intensification will increase with the economic activity of society as will the diversification of erosion and it will therefore be necessary to make use of all available knowledge and technical equipment and technologies to restrict and reduce its negative effects.

Fig. 16. Pastureland gullied by sheep trail (photo by
         D. Zachar).

2.2 CLASSIFICATION OF EROSION BY FORM

The forms of erosion are derived from the effects of exogenous agents on the soil surface — surface erosion, and under the soil surface — sub-surface erosion.

Fig. 17. Erosion of pastureland in Yugoslavia (photo by
          J. Říha).

Fig. 18.   Roadbank erosion (photo by courtesy of the Soil
Conservation Service, USA).

2.2.1 *Surface Water Erosion*

Surface water erosion may, by the effects of water on the soil surface be classified
into:

- sheet erosion

- rill and gully erosion

- stream erosion.

Sheet erosion is characterized by the detachment and removal of soil more or less
evenly over the whole affected area. The first stage of sheet erosion is selective
erosion when surface runoff removes fine soil particles and the chemical substances
which they bind; the soil texture changes as does the soil nutrient content. Soils
exposed to selective erosion become coarse grained and have a significantly lower
nutrient content while the soils enriched by the sediments are fine grained and
rich in soil nutrients.

Selective erosion is a slow process, very often inconspicuous and without any visible
traces. It becomes obvious when fine grained material accumulates in the lower

parts of the slopes, namely after heavy rain (Fig. 19). This fine grained material
often silts ditches and communications.

Fig. 19. Lower part of slope affected by selective sheet
          erosion (photo by J. Dvořák).

Selective sheet erosion causes an uneven development of the vegetation which is
manifest in its growth, colour and quality on those parts of the slope which have
been exposed to the washing of fine grained soil particles and nutrients and in
the lower parts of slopes where washed fine-grained soil particles have accumulated.
Selective erosion may reliably be ascertained by the texture analysis of the soil
and by the measurement of changes in the soil nutrient content in different parts
of the slope.

The removal of soil in thin layers (Fig. 20) is caused by runoff with a higher
kinetic energy and the unfavourable formation of the soil profile, i.e., the alter-
nation of more and less resistant layers. This type of soil wash is evident either
evenly on the whole slope or in broad strips depending on the soil surface relief.
It usually results in the total loss of topsoil.

Concentrated surface runoff rills and grooves the soil surface and gradually cuts
deep gullies into the soil surface.

In the first stage small narrow channels are cut into the soil surface which create
a dense network on the affected slope (Fig. 21a) or shallow wider channels which
are not so dense (Fig. 21b). Continuing erosion and the further concentration of

surface runoff cuts deeper channels into the soil surface, i.e., rills, which
gradually merge and deepen (Fig. 22) and become gullies. Gully erosion (Fig. 23)
may then gradually develop into dangerous and devastating ravine erosion.

Fig. 20. Layer sheet erosion (photo by J. Dvořák).

There are basically two types of gullies. In areas where the subsoil layers and
the bedrock are more resistant to the effects of water than are the surface layers,
gullies and ravines are formed with sloping sides and V-shaped bottoms. In areas
with equally resistant layers in the whole soil profile, e.g., in alluvial clays
or aeolian loess soils, gullies and ravines are formed which have vertical sides
with broad U-shaped bottoms (Figs. 24 a,b,c,d,e and Fig. 25). Water flowing over
the heads and sides of gullies and ravines is termed waterfall erosion and it
often deepens the channel at the foot of the waterfall and the banks are undermined

and caved in. Gullies and ravines often affect aquifers reducing the groundwater level and drying up the surrounding area.

Fig. 21a. Field affected by rill erosion (photo by M. Holý)

Stream erosion is caused by the stream in water courses. The stream will either erode the bed or the banks of the stream channel (Figs. 26 a, b and Fig. 27). Bed erosion is a form of longitudinal erosion proceeding along the longitudinal axis of the water flow while bank erosion is a form of transversal erosion proceeding normally to the flow axis.

Stream erosion is most evident in torrents which usually carry large amounts of sediments.

Fig. 21b. Field affected by rill erosion (photo by M. Holý)

2.2.2 Subsurface Water Erosion

This term is sometimes used to describe the transport of soil particles and nutrients
from the surface to the lower soil layers with infiltrating precipitation. This
process, however, belongs among the normal soil forming processes and should therefore not be classified as erosion.

In soils which are easily exposed to the effects of water erosion, especially
loess, groundwater scours the subsoil which it disturbs and accumulates on the
impermeable layer. The tunnels which are thus formed reduce the stability of the
overburden. In many cases the tunnel roofs fall in thereby forming deep gullies.
Tunnel erosion is therefore often classified as gully erosion. In karst areas,
karst erosion is very frequent and widespread.

2.2.3 Wind Erosion

Wind erosion differs from water erosion in that it affects large surfaces and only
in unique cases does it affect the land in strips in the prevailing direction of
the wind current.

There are two basic types of wind erosion depending on whether soil particles are
blown away — this is known as deflation, or whether solid rock formations are

ground off by sharp soil particles which are carried by the wind — this type of
erosion is known as corrasion.

Fig. 22. Higher stage of rill erosion (photo by J. Říha).

By the deflation the soil particles are transported and deposited at various
distances and often form sand dunes especially on the sea coast and in inland
deserts.

Fig. 23.   Gully erosion in Stewart County, Lumkin, USA (photo
by courtesy of the Soil Conservation Service, USA).

Corrasion is the abrasion of rocks by soil particles which are deflated by the wind.
Its intensity is determined by the resistance of the rock formation, the type and
shape of the wind borne soil particles and wind velocity.   The least resistant
rocks to this type of erosion are rocks that are easy to cut, such as sandstone.

Areas affected by corrasion have a specific character.   Corrasion shapes the rocks
into characteristic formations, such as pillars, galleries, rock towns (Fig. 28)
or embedded boulders (rocking stones) (Fig. 29).   In the Central Asian deserts a
wellknown formation are the so-called jardangs — parallel trenches separated by
mounds caused by wind corrasion.   S. Hedin[11] writes that a trench 6 m deep will
take 1600 years to cut at an intensity of 4 mm per year.

Fig. 24 a. Origination of rill.   Fig. 24 b. Development of rill.

2.3 CLASSIFICATION OF EROSION BY INTENSITY

Erosion intensity is usually expressed by the detachment and transport of soil in
weight or volume units, in some cases by the thickness of the layers of transported
matter per unit of area per unit of time. The intensity of gully erosion is often
measured by the density of rills, gullies and ravines expressed by their length
per unit of area.

By its intensity erosion may be classified as normal and abnormal, i.e., accelerated

In normal erosion processes intensity is low and loss of soil particles is offset
by the formation of new soil particles from the paternal substrate. The thickness
of the soil profile is not reduced and the texture of the surface soil changes and
becomes course grained.

Normal erosion includes seasonal erosion, manifest when soil is covered with crops
that give small protective cover and microerosion which causes the detachment of
soil particles and plant nutrients from local elevations and their transportation
to smaller distances. Seasonal erosion is manifest by a temporary decrease in soil
fertility, microerosion by a varied yield of crops.

The wash of soil particles caused by accelerated erosion is such that there can no
longer be any natural soil replacement from the paternal substrate. The result is
a sharply modelled surface.

Fig. 24 c. Development of rill   Fig. 24 d. Development of gully
            into gully.   into ravine.

It is extremely difficult to determine the permissible erosion rate which is given
by erosion intensity because it must always be considered with regard to the possible consequences of the erosion process in the given conditions.

In agriculture the permissible erosion limit is considered as being that intensity
which allows the formation of new soil and does not reduce current soil fertility.

H. H. Bennet[2] assumes that a 2-3 cm soil layer will take 200-1000 years to build
of the paternal substrate providing it has good vegetative cover and effective
soil protection. Under the given conditions a layer 0.026-0.13 mm in thickness
would be formed per year which is 324-1620 kg ha^{-1}.

Z. Kukal[7] has derived from data published in literature on the intensity of soil
weathering under diverse conditions that the average rate of soil formation on
the Earth's surface is approximately 10 cm in 1000 years which is 0.1 mm per year,
i.e., 1 m^{3}ha^{-1}.

R. M. Smith and W. L. Starney[8] studied the intensity of the erosion process on
experimental plots with virgin soils protected by vegetative cover. Measurements
which they conducted in different parts of the USA showed that normal erosion was
within the region of 0.25-1.48 t ha^{-1} per year which corresponds to natural soil
formation.

Fig. 24 e.   Ravine with dejection cone (photo by M. Holý).

The permissible erosion rate is determined on the basis of considerations on the
soil formation rate and the economics of farm production. The permissible erosion
limit differs with the character of the soil and the depth of the soil profile.
The aim is not to allow the soil loss to exceed 1.25 t ha^{-1} per year[5]. In the
US the permissible erosion rate is within the region of 2.5-12.5 ha^{-1} per year.
In Czechoslovakia the permissible erosion rate of shallow soils (soil layer up to
30 cm in thickness) is 1 t ha^{-1} per year; in soils with a thickness of 30-60 cm the
permissible rate is 4 t ha^{-1} per year; in deep soils (soil layer 60-120 cm in
thickness) the permissible rate is 10 t ha^{-1} per year; in very deep soils with a
soil layer of more than 120 cm in thickness on loess and clay slopes the permissible
erosion rate will be 16 t ha^{-1} per year.

The said permissible soil loss tolerance is derived with the aim of preserving
soil fertility. Currently, when soil erosion is becoming a serious source of
the pollution of water resources the permissible erosion rate should be determined
with regard to the detachment and transport of chemicals and the required quality
of water resources in the given area.

REFERENCES

1. Beasley, R. P., *Erosion and Sediment Control*, Iowa, 1972.
2. Bennet, H. H., *Elements of Soil Conservation*, New York-Toronto-London, 1955.
3. Economic Commission for Europe, *Proceedings, Seminar on the Pollution of Waters
      by Agriculture and Forestry*, Vienna, 1973.

4. Hagan, R. M., Haise, H. R. and Edminster, T. W., *Irrigation of Agricultural Lands*, Wisconsin, 1967.
5. Hudson, N., *Soil Conservation*, London, 1973.
6. Konke, G. and Bertran, A., *Okhrana pochvy*, Selchoz-literatura, Moskva, 1962.
7. Kukal, Z., *Geology of Recent Sediments*, Prague, 1964.
8. Smith, R. M. and Starney, W. L., Determining the Range of Tolerable Erosion, *Soil Sci. Proc.* 6, 1965.
9. Štěpa, B., *Melioratsiya zemyel v SSSR*, Moskva, 1975.
10. Vice, R. B., Guy, H. P. and Ferguson, C. E., Sediment Movement in an Area of Suburban Highway Construction, *Geological Survey Water-Supply Paper* 1591-E, 1969.
11. Zachar, D., *Soil Erosion*, SAV, Bratislava, 1970.

Fig. 25.   Farmland devastated by ravines (photo by courtesy of the Soil Conservation Service, USA).

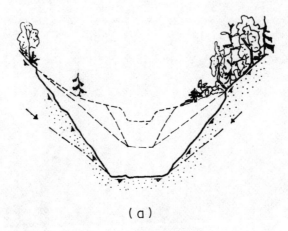

(a)

Fig. 26a. Stream bed erosion.

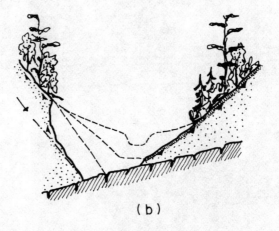

(b)

Fig. 26b. Stream bank erosion.

Fig. 27. Stream bank erosion (photo by J. Říha).

Fig. 28.   Rock formation in Bohemian Central Mountain Range
           (photo by J. Říha).

Fig. 29.   Rock formation near Prague (photo by J. Říha).

3. Mechanism of Erosion Processes

Erosion is a process which is continually transforming the Earth's surface. It is initiated by natural forces and intensified by human activity which has been significant namely in the recent period, i.e., since man began to step up the exploitation of natural resources for economic development.

The mechanism of erosion processes is initiated and controlled by the interaction and action of a wide range of factors, the most prominent being:

- climatic and hydrological agents
- morphological agents
- geological and soil agents
- vegetative agents
- technical agents
- socio-economic agents.

3.1 CLIMATIC AND HYDROLOGICAL AGENTS

The climatic and hydrological conditions of an area are characterized by geographical position, altitude, atmospheric temperature, precipitation, evaporation, air humidity, direction and force of wind currents and surface runoff. For erosion control it is necessary to investigate namely the occurrence, distribution and intensity of precipitation and the formation and course of surface runoff.

3.1.1 Precipitation and its Physical Characteristics

Any investigation of erosion processes should include an evaluation of atmospheric precipitation with regard to its direct effect on the soil surface and on the runoff caused thereby. The erosion effects of liquid and solid precipitation differ. The erosion effects of heavy rainfall causing high intensity surface runoff are intensified by the effects of the kinetic energy of raindrops on the soil surface. The raindrops falling on the soil surface cause a splatter of soil particles in the soil aggregates which are then carried away by surface runoff.

With regard to water erosion the most unfavourable rainfall in the temperate climatic zone are heavy rains with a duration of up to 180 minutes and a level of 10-80 mm[26];

38

in the tropical zone the intensity of storm rain is higher - after N. Hudson[31]
it is up to 200 mm per hour and duration is usually more than 120 minutes.

The rainfall intensity curve may be expressed as function

$$i = \frac{A}{(B + T)a} \qquad (3.1)$$

where i is rainfall intensity (mm min^{-1})

  T is rainfall duration (mins)

  A,B,a are parameters of the rain gauge station.

The values of theoretical rainfall intensities are given in Table 3.1 after J.
Haeuser (maximal)[24] and O. Dub (engineering)[16].

TABLE 3.1   Theoretical intensities of heavy rainfalls

Rainfall duration (min)	5	10	15	20	30	60	90	100
Maximum intensity (mm min^{-1})	7.00	5.40	4.47	3.84	3.07	2.08	1.64	1.38
Engineering intensity (mm min^{-1}) after O. Dub	5.40	3.82	3.06	2.59	1.95	1.20	0.87	0.72

A number of empirical equations have been written for expressing the relation
between the intensity of heavy rainfall and its duration.

Significant is the relation derived by G. A. Alexeyev[1] from a large number of
statistical data. The result gave the expression

$$i_{max} = \frac{A + B \log N}{(1 + T)^{2/3}} \qquad (3.2)$$

where i_{max} is maximum rainfall intensity (mm min^{-1})

  T   is rainfall duration (mins)

  A,B   are geographical parameters dependent on the climatic conditions of the
    area

  N   is the probable number of years in which rainfall at intensity i and
    duration T will occur only once.

G. A. Alexeyev found that the relation between rainfall intensities with a different
periodicity is fairly constant for a given area regardless of rainfall duration.
This has also been confirmed by M. Dzubák[19].

The intensity of heavy rainfall decreases with the size of the affected area. This
inverse relation is in agreement with the observations of other authors including
J. Haeuser[24] (Fig. 30).

Rain consists of raindrops varying in size. Measurements made by N. Hudson[31] and
D. C. Blanchard[8] have proved that the biggest raindrops are 5 mm in diameter.
Raindrops which are > 5 mm in diameter may split into smaller drops. Regional low
intensity precipitation is usually made up of small drops, while high intensity
rainfall is usually characterized by drops of much bigger diameter.

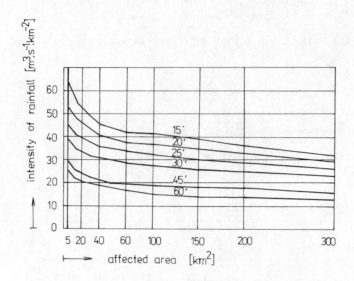

Fig. 30.   Relations between the intensity of heavy rainfall
            and the area of the affected catchment after
            J. Hauser.

It is extremely difficult to describe raindrop distribution by a single parameter.
N. Hudson[32] states that probably the best index for drop distribution is the
median volume drop diameter D_{50} which is such that half of the volume of the rain
falls in drops with a smaller diameter and the other half as bigger drops.   The
index is obtained from a plot of cumulative volume against drop diameter (Fig. 31).

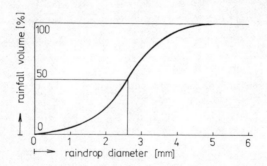

Fig. 31.   Relation between the size of raindrops and rain-
            fall volume.

With regard to the upper limit of drop size this index only applies to low intensity
rainfall.

The relation between raindrop size and rainfall intensity is given by V. N. Obolenskiy
and V. Y. Nikandrov[43] and by N. Hudson[32] and is shown in Fig. 32.   The relation
given by N. Hudson clearly shows that the drop median volume diameter decreases at
very high intensities.

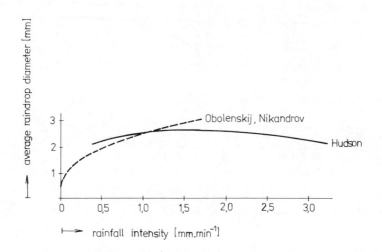

Fig. 32.   Relation between the average diameter of raindrops
            and rainfall intensity.

The velocity of falling drops of rain is affected by gravity and by resistance of
the air.   The raindrop falls freely under the force of gravity and will accelerate
until the frictional resistance of the air equals the gravitational force and will
then continue to fall at a constant speed.   This terminal velocity depends on the
size and shape of the raindrop.

The terminal velocity of the raindrops has been studied by many authors.   The relation between velocity and drop diameter was studied by J. O. Laws[40] using high-
speed photography.   The results are shown in Fig. 33.

The accuracy of these results was confirmed by R. Gunn and G. D. Kinzer[23] who
gave the drops a slight electric charge and allowed them to fall through induction
rings which gave an electric impulse when the drop fell through them.   The impulses
were fed through amplifiers to an oscillograph so that the time between the two
induction rings could be measured very accurately.   The data differed from those
obtained by J. O. Laws by a mere 3%.

V. V. Slastikhin[54] gives the following expression for the calculation of raindrop
velocity v_k

$$v_k = 13\sqrt{d} \ [\ ms^{-1}\ ] \qquad\qquad (3.3)$$

where d is the drop diameter (cm).

When rain is accompanied by wind there is the added component of velocity and the
resultant vector may be greater than still-air velocity.   The effects of wind are
more significant in regional rainfall with raindrops of smaller diameters than in
torrential rainfall with bigger drops.   J. O. Laws[40] observes that even in torrential rainfall accompanied by wind approximately 95% of the raindrops will
impinge on the soil surface at the same terminal velocity as in still air.

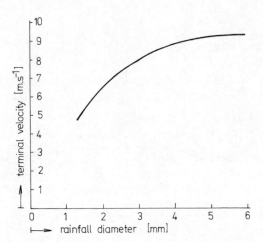

Fig. 33.   Relation between raindrop velocity and raindrop
diameter after J. O. Laws.

The kinetic energy of raindrops has fundamental importance for the erosion process.
Raindrops impinging on the soil surface break down the soil aggregates and detach
the soil particles splashing them to short distances and increasing the turbulence
of surface runoff (Figs. 34 a and b).

It is very difficult to conduct direct measurements of the kinetic energy of rainfall
because the forces involved are so small that it is very difficult to design an
instrument sufficiently sensitive and one that would not be swamped by the effects
of wind. It seems to be more convenient to convert kinetic energy into another
form of energy, one that may be more easily measurable in small values. Thus, for
instance, it is possible to measure the momentum of raindrops falling on diaphragm
which emits sound. The momentum is measured by the intensity of the sound emitted
by the falling raindrops.

Usually, however, the momentum of the falling rain is calculated from the raindrop
mass and their terminal velocity. The kinetic energy of rainfall is the summation
of the kinetic energy of the individual raindrops.

The relations between kinetic energy of rainfall and its intensity as given by
various authors are given in Fig. 35.

3.1.2 Erosivity of Rainfall

Erosivity is the potential ability of rain to cause erosion. It is a function of
the physical characteristics of rainfall.

The significance of certain characteristics of rainfall and their relation to the
intensity of the erosion process, most frequently the dependence of the erosion
process to rainfall intensity, has been emphasized by many authors, namely T. H.
Neal[45], G. Dakshinamurti and T. D. Biswas[13], G. András[3], H. H. Krusekopf[36],
H. Kuron, L. Jung and H. Schreiber[37] and others.

T. H. Neal who studied the relation between rainfall intensity and soil wash on
soil monoliths with different gradients gives the expression

$$S_p = KI^{0.8}i^{1.2} \tag{3.4}$$

where S_p is soil wash (t ha^{-1})

    I is the slope gradient in degrees

    i is rainfall intensity (mm min^{-1})

    K is the coefficient dependent on local conditions.

Fig. 34 a. Impact of a raindrop on soil (photo by courtesy
of the Soil Conservation Service, USA).

The given relations for slope gradients between 0 to 14.4° and rainfall intensity from 0.38 to 1.67 mm min^{-1} are plotted in Fig. 36.

M. Holý[29] investigated the relation between the intensity of torrential rainfall and soil wash by simulating rain in a rain simulator over a tilting hydraulic flume with natural soil (Figs. 37 and 38). He gives the following expression for the relation where x is soil wash (kg ha^{-1})

$$x = a.y^b \tag{3.5}$$

where y is a rain intensity (1 s^{-1} ha^{-1})

a is the coefficient dependent on soil properties

b is the exponent dependent on the slope gradient.

Fig. 34 b.   Impact of raindrops on soil (photo by J. Říha).

The investigated relations for slope lengths 8.55 m and 17.10 m are plotted in Figs. 39 and 40.

The most significant characteristic of rainfall is the kinetic energy of raindrops impinging on the soil surface which is considerably higher than the energy of surface runoff. This is evident from comparisons of the kinetic energy of rainfall and of surface runoff.

Let us assume the mass of raindrops m of which 50% is involved in surface runoff owing to infiltration and the retention capacity of the soil surface. Raindrops with an average size of 3 mm have an impact velocity of 8 m s^{-1} (Fig. 33); the velocity of the surface runoff is assumed at 1 m s^{-1}. Under the given conditions the kinetic energy of the rainfall is $E = \frac{1}{2} m \, 8^2 = 32 \, m$, the kinetic energy of surface runoff is $E = \frac{1}{2} \frac{m}{2} \, 1^2 = \frac{m}{4}$. The kinetic energy of the rainfall is in this case 128 times bigger than the kinetic energy of the surface runoff.

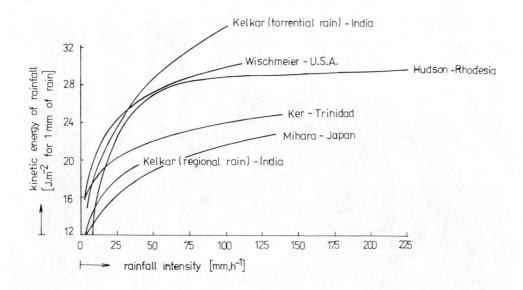

Fig. 35. Relation between rainfall kinetic energy and rainfall intensity.

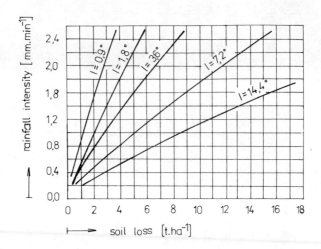

Fig. 36. Relation between rainfall intensity and soil erosion after T. H. Neal.

Each of the given energies is involved in the erosion process in a different way. The kinetic energy of rainfall breaks down soil aggregates, the kinetic energy of the surface runoff not only breaks down the soil aggregates but also detaches and transports them. Experiments[32] have shown that the significant reduction of the kinetic energy of rainfall at the same surface runoff led to a significant decrease in soil splash.

Fig. 37.   Hydraulic flume with rain simulator (photo by
           courtesy of the Institute of Irrigation and Drain-
           age, Technical University, Prague).

The relation between soil wash, mass and the terminal velocity of raindrops which
characterize their kinetic energy has been studied by many authors. W. D. Ellison[20]
for instance, gives the following expression

$$S_p = 93.56 \ K(v_k^{4.33} \ d^{1.07} \ i^{0.65})$$
                                                                    (3.6)

where S_p is soil wash (g per 30 mins)

   v_k is drop velocity (m s^{-1})

   d   is drop diameter (mm)

   i   is rainfall rate in (cm per hour)

   K   is dependent on soil conditions.

The dependence of soil wash on the kinetic energy of rainfall expressed by other
authors on the basis of laboratory experiments was confirmed by W. H. Wischmeier[60]
for natural rainfall. After processing 8250 data from 35 experimental stations in
the USA he arrived at the conclusion that the most convenient estimator of the

relation between rainfall and soil loss is a parameter expressing the product of
the kinetic energy of the rainfall and its maximum 30 minutes intensity (Fig. 41).

Fig. 38. Sampling of silt (photo by courtesy of the Institute
          of Irrigation and Drainage, Technical University,
          Prague).

The said parameter of erosivity was described as the EI_{30} Index. It can be computed
for individual rainfalls and the values can be summed up over periods of time to
give daily, weekly, monthly, annual or n-year values of the parameter. For its
computation W. H. Wischmeier considered rainfall levels ≥ 12.7 mm.

The EI_{30} Index which has been proven in the US can be applied in other conditions
only with certain modifications. For conditions prevailing in Africa, N. Hudson[32]
introduced the KE Index being the total kinetic energy of the rain falling at intensities of more than 25 mm per hour.

In Czechoslovakia the values of the EI_{30} Index were investigated by J. Pretl[50]
who found that there exists a close correlation between the mean annual value of
EI_{30} and the average annual total precipitation. For Bohemia he expressed this
relation as

$$EI_{30} = 0.679H_r + 4.279\ 3$$
$$H_r = 10.156EI_{30} + 183.347 \qquad\qquad (3.7)$$

with the coefficient of correlation $r_{x,y} = 0.83$.

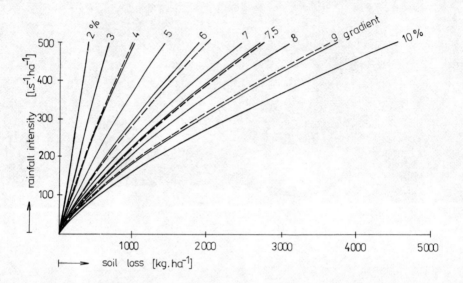

Fig. 39. Relation between rainfall intensity and water
erosion for slope length 8.55 m.

3.1.3 *Solid Precipitation*

Solid precipitation significant for erosion is snowfall because in the spring thaw
it may cause considerable surface runoff.

Runoff from snowfall is dependent on the physical properties of the snow and on
the depth and distribution of the snow cover.

The density of the snow changes owing to gravity, solar radiation and the recrystal-
lization of snow flakes. Fresh snow usually has a density of 0.02 to 0.07 g cm^{-3},
snow pack 0.20 to 0.40 g cm^{-3} and over-ripe snow 0.25 to 0.50 g cm^{-3}.

The value of snow also indicates the value of water which is expressed as the ratio
of the volume of snow to the volume of water resulted from the melting of the snow
(dimensionless number) or as the ratio of weight to the volume of snow (g cm^{-3}).

The water capacity of snow (%) expressing the capacity of snow to retain a certain
amount of water in liquid state depends mainly on its grain size. In fresh snow
it is 35-52%, in fine-grained and medium-grained snow 25-35%, in course-grained
snow 15-25%.

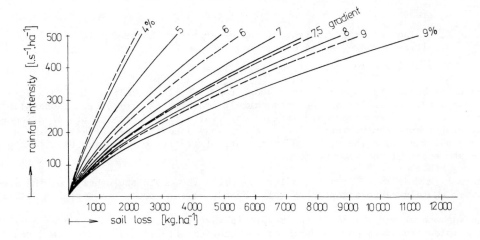

Fig. 40. Relation between rainfall intensity and water ero-
          sion for slope length 17.10 m.

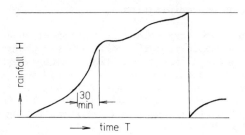

Fig. 41. Determination of maximum 30 mins rainfall intensity
          from rain gauge record.

The distribution of the snow cover depends on the precipitation and temperature
conditions of the given area. The snow cover is formed in places where the temperature of the ground layer of air remains below $0°C$ for a certain period of time, the
temperature of the soil surface does not rise above $0°C$ and the heat increase – heat
loss balance is negative.

The duration of the snow cover depends mainly on altitude and geographical position.
The dependence may be expressed as

$$D = D_O + K'H \qquad\qquad\qquad (3.8)$$

where D are days with a snow cover on an area at altitude H

   D_O are days with a snow cover at altitude H=0

   K' is the vertical gradient of the duration of the snow cover per 1 m
      altitude above sea level (day m^{-1})

   H is altitude above sea level (m).

The distribution of the snow cover shows considerable anomalies which are effected
by the wind, the relief of the area, namely its vertical relief (ruggedness), the
forest cover, of anthropogenic factors especially the effects of urbanization and
industrial emissions. There are, however, many other factors affecting the snow
cover and therefore the method of mathematical statistics is being used to express
the character of the snow pattern.

There is a wide range of methods for measuring the depth and density of the snow
cover. Lately radioisotopes have also been used for ground and aerial monitoring.

The snow cover and the resulting runoff in the spring thaw in a given area will be
calculated from data on the depth of the snow cover, its distribution and density.

3.1.4 Runoff

Surface runoff transports soil particles detached by raindrops and by its tangential
stress it detaches further soil particles and chemical substances for transportation

3.1.4.1 Runoff from Liquid Precipitation

Surface runoff from the slope takes place at the moment when rainfall intensity
exceeds the infiltration capacity of the soil. This depends on many factors of
which the most important are climatic conditions, topography, physical factors,
i.e., the morphology of the area, the geographical and soil conditions and the
composition of the vegetative cover, and the anthropogenic factors which have
unfavourable effects on the water regime of the area. The infiltration capacity
of the soil decreases with time insomuch as the infiltrating water fills the soil
pores until it reaches a more or less constant value.

In most cases runoff from torrential rainfall is of major importance for erosion
control measures. This runoff may basically be studied using three methods:

- using the runoff coefficient from torrential rainfall of average intensity

- using empirical relationships derived from observations and field measurements

- by the evaluation of rainfall from the relation between rainfall, infiltration
   and intensity and duration of the rainfall.

The determination of rainfall from torrential rainfall using the runoff coefficient
requires that the correct value of the runoff coefficient be determined such as
would express the ratio of the total rainfall and runoff. To determine this ratio
it is necessary to compare torrential rainfalls and the respective runoffs and to
evaluate them in different conditions. To simplify calculations only the most
important and most easily measurable factors affecting runoff are usually considered,
namely the character of the area, its relief, soil properties and vegetal cover.

O. Härtel[25] writes the following expression for determining the runoff coefficient

$$o = n_1 \ n_2 \ n_3 \ n_4 \tag{3.9}$$

where o is the runoff coefficient

n_1 is the coefficient expressing the length of the valley affected by rainfall

n_2 is the coefficient expressing the effects of the forest cover

n_3 is the coefficient expressing the relief

n_4 is the coefficient expressing the effects of the soil permeability.

Coefficients n_1 to n_4 after Härtel are given in Table 3.2.

TABLE 3.2.   Partial runoff coefficients after O. Härtel

Length of valley	0.2	0.3	0.4	up to 10 km
n_1	0.9	0.85	0.8	0.55
Forest cover	100	75	50 25	0 %
n_2	0.6	0.7	0.8 0.9	0.95
Relief	very steep		hilly	rolling
n_3	0.90		0.85	0.80
Soil	imper- meable	little permeable	medium permeable	highly permeable
n_4	0.90	0.80	0.75	0.70

A. N. Kostyakov[34] derived the runoff coefficient for different reliefs and soil permeability from field measurements.   The coefficients are shown in Table 3.3.

TABLE 3.3   Runoff coefficients after A. N. Kostjakov

Soils of catchment area	Slope degree of catchment		
	small < 0.01	medium 0.01-0.05	big > 0.05
Highly permeable	0.10-0.20	0.15-0.25	0.20-0.30
Semi-permeable	0.15-0.25	0.20-0.30	0.25-0.40
With medium permeability	0.20-0.30	0.25-0.45	0.35-0.60
Low permeable	0.25-0.40	0.30-0.60	0.50-0.75
Frozen	0.35-0.60	0.40-0.75	0.80-0.95

The runoff coefficient was studied by a number of other authors such as F. Wang[59], J. Schönfeldt[52], A. Hoffman[27], G. Förster[22], C. E. Ramser[51], H. L. Cook (cit.32) and others.   The applicability of their methods should be tested for the investigated conditions.

A more detailed study of the effects of different factors on the runoff coefficient shows its significant dependence on the duration of rainfall.

W. C. Hoad (cit. 12) expresses this as

$$o = \frac{aT}{b + T}$$

(3.10)

where T is rainfall duration (mins)

a,b are constants dependent on soil permeability.

L. G. Dyemidov[14] uses the following expression

$$o = \mu \, i_s^x \, T^y$$

(3.11)

where i_s is the mean rainfall intensity (1 s^{-1} ha^{-1})

T is rainfall duration (mins)

μ, x, y are constants dependent on local conditions.

The dependence of the runoff coefficient on raintall duration for soils of various permeability is given by G. M. Fair[21] (Fig.42).

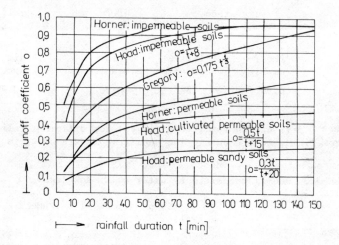

Fig. 42. Dependence of runoff coefficient on rainfall
          duration after G. M. Fair.

The surface runoff may be derived from known data on rainfall intensity, runoff coefficient and the area of the catchment using the expression

$$Q = oiP$$

(3.12)

where Q is surface runoff (m^3 s^{-1})

i is the intensity of the considered rainfall (m s^{-1})

o is the runoff coefficient

P is the area of the catchment (m^2).

The application of expressions (3.10 and 3.11) assumes that runoff occurs during rainfall when T' = T (T' = runoff duration and T = rainfall duration). Such runoff, however, only occurs in very small catchment areas with steep sloping land in high-intensity torrential rainfall. In large catchments a retardation of run-off occurs resulting from the soil surface retention capacity; rainfall duration is shorter than runoff duration T<T'.

The time needed for the water particles to reach the outlet profile from the hydraulically most distant point in the catchment, described as runoff concentration time T_k is affected by the area of the catchment, its topography and relief.

In order to make approximate determinations of concentration time T_k it is possible to use data provided by Q. C. Ayres[5] (Table 3.4), R. M. M. Cormack[11] plotted in graph shown in Fig. 43.

TABLE 3.4   Runoff concentration time after Q. C. Ayres

Area of catchment (ha)	Concentration time (mins)
0.4	1.4
2.0	3.5
4.0	4.0
40.5	17.0
202.5	41.0
405.0	75.0

Proceeding from measured runoff concentration time T_k for the given conditions it is then possible to select the intensity of torrential rainfall for duration T = T_k and to use relation (3.12) for computing the runoff value.

The determination of the values of surface runoff from empirical relations written on the basis of observations and field measurements is given by the existence of these relations. Specially important for erosion control are relations derived from investigations of the effect on runoff of torrential rainfall and on natural conditions. These relations are given by a group of formulas for the computation of the runoff from a small catchment, i.e., such a catchment in which maximal run-off is caused by torrential rainfall.

The general formula for computing the maximal runoff from a small catchment area may be written

$$Q_{max} = iP\phi \qquad (3.13)$$

where Q is maximal runoff

i is rainfall intensity resulting in maximal runoff

P is the area of the catchment

ϕ is the runoff coefficient of maximal runoff.

Well-known is the formula written by A. Specht[55]

$$q_{max} = i\,\frac{10}{36}\,\sqrt[12]{\frac{1}{P}}\left(0.2 + \frac{0.8}{\sqrt[4]{T}+1}\right) \qquad (3.14)$$

where q_{max} is maximal specific runoff (m^3 s^{-1} km^{-2})

    i   is the intensity of considered rainfall (mm per hour)

    P   is the area of the catchment (km^2)

    T   is the duration of the considered rainfall (hrs).

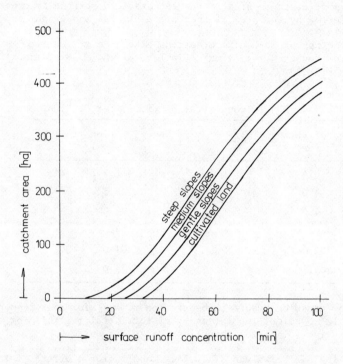

Fig. 43. Determination of surface runoff concentration
after R. M. Cormack.

The formula written by D. L. Sokolovskiy (cit.17) for small flows with catchments
without water reservoirs is written

$$Q_{max} = \frac{0.28 \, H_s \beta P f}{t}$$

(3.15)

where Q_{max} is maximal runoff (m^3 s^{-1})

    H_s is total precipitation for computed rainfall duration

         $T = t(t+1)^{-0.2}$ (mm)

    β volume runoff coefficient derived by analogy

    P   area of catchment (km^2)

    f   coefficient of the hydrogram shape whose average value is 0.6

    t   is runoff duration $t = \frac{L}{3.6v}$

L   is the length of the flow (km)

v   is the average runoff rate ($m\ s^{-1}$).

The given empirical formulas for the computation of runoff and a number of other formulas, such as those written by B. V. Polyakov[47], V. A. Ogievskiy (cit. 16), N.I. Sus[56] etc. require the knowledge of the values of various coefficients affected by local conditions.

A suitable formula for computing maximal runoff from a small catchment area has been derived in Czechoslovakia by O. Dub[16]

$$q_{max} = \frac{A_0}{(P + 1)^n} (1 + \sum_{i=1}^{4} 0_i) \qquad (3.16)$$

where q_{max} is the maximal specific runoff ($m^3\ s^{-1}\ km^{-2}$)

P   is the area of the catchment (km^2)

A_0,n are parameters expressing the relation between the maximal runoff and regional conditions

0_i expresses local effects.

Parameters A_0, n were investigated by O. Dub for local conditions.

The effects of afforestation are only considered as a deviation from the average state which is characterized by a 40 to 50% forest cover.

$+0_1$ is the relative increase in maximum runoff in floods affecting areas with a forest cover of less than 50% and -0_2 is the decrease in maximum runoff in floods affecting areas with a forest cover of more than 50%. The effects of the shape of the catchment area is given for fan-shaped catchment areas. O. Dub gives the value expressed as $0_3 = 0.05$ to 0.1, for elongated catchment areas the value is $0_4 = -0.1$.

In order to evaluate the runoff values from empirical expressions it is useful to conduct an informative comparison with values measured in similar conditions. The range of specific runoff q_{100} in dependence on the area of the catchment for a catchment area of up to 100 km^2 is given in Fig. 44 (cit.10). The shape and forest cover of the catchment was not considered.

In a situation where direct observations have not been made, empirical expressions will be found suitable for ascertaining maximum runoff from small catchments. They give the absolute maximum runoff or runoff with very low periodicity (usually p = 0.01) which is important for river training. With regard to erosion control on agricultural soils the obtained values are excessively high and measures suggested on such a basis would in most cases not be economic. From the known Q_{100} (runoff occurring once in 100 years) it is possible to derive water with different periodicity of occurrence. The relation between n-year water Q_n and Q_{100} may be expressed as

$$Q_n = \alpha_n\ Q_{100} \qquad (3.17)$$

Values derived in Czechoslovakia[9,15] for a catchment area of up to 30 km^2 (Table 3.5) may be useful for determining the value of coefficient α_n.

Runoff may be determined from the evaluation of the dependence of infiltration on rainfall intensity and duration provided that rain recording gauge measurements are made and that the course of the infiltration of rainfall into the soil is expressed.

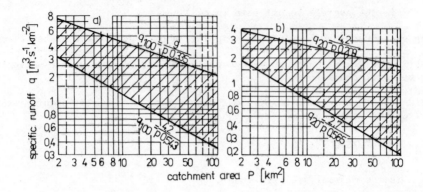

Fig. 44.   Relation between q_{100} and area of catchment after
           M. Čermák.

TABLE 3.5   Values of coefficient α_n after A. Bratránek and O. Dub

N years	Steep terrain with no forest cover (extreme conditions) after O. Dub	Hilly terrain (with 30–60% of the area under forest cover) after A. Bratránek	Rolling terrain (with 60–80% of the area under forest cover) after A. Bratránek	Lowland with parts of the area under forest cover (extreme) after A. Bratránek
1	0.06	0.10	0.14	0.18
2	0.08	0.15	0.21	0.29
5	0.13	0.23	0.33	0.44
10	0.21	0.33	0.45	0.55
20	0.34	0.47	0.60	0.67
50	0.62	0.70	0.81	0.84
100	1.00	1.00	1.00	1.00

The course of the infiltration of rainfall into the soil is usually measured by
various infiltrometers or by computations applying various formulas, e.g., N. A.
Kostyakov's expression[34], V. S. Mezentsev's expression[42], J. R. Phillip's
expression[46], etc.

With regard to erosion control the course of the infiltration of rainfall into the
soil was measured in Czechoslovakia for various types of soils by simulating rain
on experimental plots 2m x 1m in area. For determining the soil infiltration
capacity the following expression was used[18]

$$V_t = v_c t + \frac{v_1 - v_c}{1 - \alpha} \, t^{(1-\alpha)} \tag{3.18}$$

where V_t is total infiltration of water into the soil (mm)

   v_c soil permeability (mm min^{-1})

   t time (mins)

v_1 infiltration of water into the soil in time t=1 (mm min^{-1})

α constant dependent on soil properties.

Sprinkler intensity was equalled to the infiltration capacity of the soil and the time values v_1 and α were thus obtained. Soil permeability v_c was obtained by calculating for each trial the instantaneous infiltration values for 5 min intervals from the measured values of total infiltration; median values were then determined from these.

The value of the soil infiltration capacity was then averaged for t≥60 mins.

The dependence of the instantaneous infiltration of water v_t on time t yielded the derivation of the expression (3.18). The resulting expression may be written

$$v_t = v_c + (v_1 - v_c) \ t^{(-\alpha)}$$ (3.19)

which corresponds to the relation given by V. S. Mezentsev[42].

Two other values were introduced, namely the slope which was found to affect infiltration v_1 and coefficient α on runoff plots of the same area with slopes from 2 to 10°. The following expression was then written

$$v_1 = bi^\gamma$$ (3.20)

where i is rainfall intensity (mm min^{-1})

　　b is the coefficient dependent on soil properties

　　γ is the exponent dependent on soil properties.

and

$$\alpha = ci^\delta I^\varepsilon$$ (3.21)

where c is the coefficient dependent on soil properties

　　δ,ε are exponents dependent on soil properties

　　I is the slope of the runoff plot (%).

When the coefficients and exponents have been obtained for the investigated conditions these relations will allow us to obtain value v_t in time for the given soil, constant rainfall intensity i and slope I and will thus allow the plotting of the time course of the infiltration of rainfall into the soil.

M. Holý[30] measured the runoff of torrential rain by reducing total precipitation according to the infiltration curve.

The dependence of rainfall duration T (min) on rain depth h(mm) expressed by the record from the rain recording gauge was converted to the dependence of time T on specific rainfall intensity (1 s^{-1} ha^{-1}). The curve of specific rainfall intensity was converted into a "curve of elementary runoff" by deducting infiltration in time intervals at constant intensity. The problem was solved from the measured infiltration curve. It was assumed that the relation between rain intensity and infiltration is approximately maintained for all intensities important for soil conservation, e.g. about 100 1 s^{-1} ha^{-1}. It was thus possible to obtain for each interval the infiltration/runoff ratio. (In Fig. 45 — infiltration is the black coloured space, runoff is the white coloured space). M. Holý constructed a nomogram (Fig. 46) for the instantaneous identification of surface runoff intensity which makes it possible to obtain from the recordings on the rain recording gauge, divided into time intervals of equal steepness and subsequent identification of the average depth of rainfall in these time intervals, the respective coordinates of the points of the elementary runoff curve.

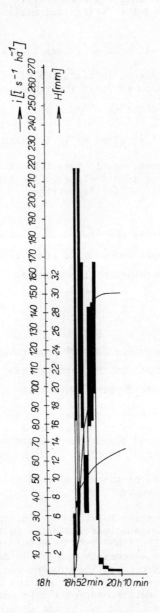

Fig. 45.   Runoff-infiltration ratio of precipitation.

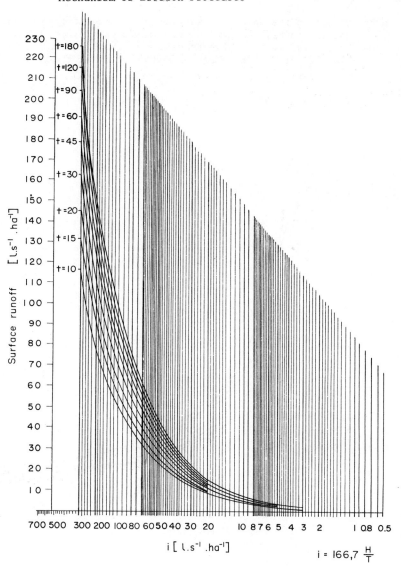

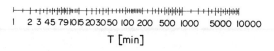

Fig. 46. Nomogram for identification surface runoff
intensity after M. Holý.

In order to obtain the values of surface runoff from a greater number of rains, a
curve of theoretical elementary runoffs should be plotted. This curve will be
obtained by determining the highest values of runoff in the elementary runoff
curve for the selected time intervals and by plotting these values as ordinates in
dependence on time. By linking these points for individual rainfalls we shall
obtain a system of lines of elementary runoffs whose envelope is the sought "line
of theoretical elementary runoff" (see Fig.47).

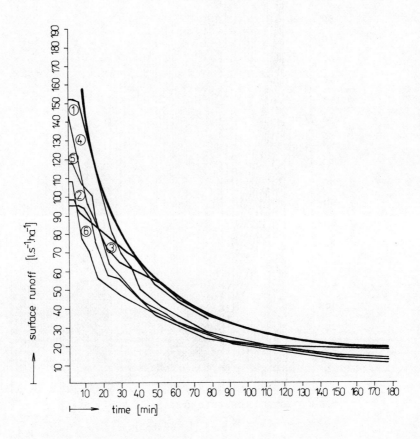

Fig. 47. Line of theoretical elementary runoffs.

A closer study of the "line of elementary runoff" made by the author showed that
it may be expressed by the equation of a hyperbola written

$$q = \frac{A}{T + B} \qquad\qquad (3.22)$$

where q is the specific runoff (1 s^{-1} ha^{-1})

     T is the rainfall duration (mins)

    A,B are coefficients in values

$$A = \frac{-(q^2 T)(q) + (qT)(q^2)}{k(q^2) - (q)(q)}$$

$$B = \frac{-k(q^2T) - (qT)(q)}{k(q^2) - (q)(q)}$$

k is the number of investigated rainfalls

( ) is the sign for the sum.

The determination of surface runoff from torrential rainfall by evaluating rainfall from the dependence of infiltration on rainfall intensity and duration requires an investigation of the infiltration of rainfall into the soil under varied conditions. It eliminates the shortcomings of determining runoff using the coefficient of runoff or empirical relations and, with certain simplifications, approaches the real course of runoff in nature.

3.1.4.2 Runoff from Solid Precipitation

With regard to erosion control, runoff from solid precipitation is usually considered as being runoff from snow. Such runoff is dangerous during the spring thaw when the wet top soil layer does not permit the snow melt to infiltrate into the soil. The runoff flowing down slopes washes away the soil slurry and this process recurs the following day. In the meantime the soil has hardened overnight owing to a drop in the night temperature. This process can result in considerable soil loss, especially on sunny slopes.

In order to be able to determine the values of runoff from snow precipitation the process of the snow thaw should be known.

The snow thaw is affected by the combined action of many agents:

- in cloudy weather with the temperature of the air rising above $0°$ C snow melts as a result of air temperature (advective type of thaw):

- in clear weather with temperatures rising above $0°$ C with solar radiation (the advective-solar type of thaw)

- in rainy weather without solar radiation snow melts by the temperature of rain water (pluvial type of thaw)

- in weather with alternating sunny and rainy weather (the solar-pluvial type of thaw)

- in sunny weather with temperatures slightly below $0°$ C the snow melts under solar radiation (solar type of thaw).

Many authors have studied the snow thaw, especially its intensity, e.g. P. P. Kuzmin [39], V. D. Komarov [33], E. G. Popov [49] and others. They mainly considered the physical factors affecting the thaw and investigated the heat balance. The heat balance is, however, difficult to determine as it requires a number of climatic measurements which are not part of the usual routine measurements carried out by meteorological stations.

The empirical expression for the snow thaw has been written in Czechoslovakia for conditions prevailing in this country [35]

$$h' = at° - b \qquad\qquad\qquad (3.23)$$

where h' is the height of the water column of snow (mm h^{-1})

    $t°$ is the average air temperature ($°$ C)

    a,b are agents dependent on the snow thaw conditions;
        for the advective snow thaw in Czechoslovakia's conditions are a = 0.5, b = 0.5, for the advective-solar type a = 0.8, b = 3.0.

The following relation applies for calculations of the advective snow thaw over a lengthy period of time:

$$h' = a\Sigma \ (+t^\circ) - nb \qquad\qquad (3.24)$$

where $\Sigma \ (+t^\circ)$ is the sum of above zero temperatures of the air $(^\circ C)$

n is the duration of above zero air temperatures in hours

The following expression has been written for the solar-advective snow thaw:

$$h' = a\Sigma(+t^\circ) - (n-m)b \qquad\qquad (3.25)$$

where n is the time interval (in hours) in which the thaw is investigated

m is the number of hours with zero and sub-zero temperatures (in $^\circ C$) in the same time interval.

REFERENCES

1. Alexejev, G. A., *Metod ustanovleniya zavisimosti mezhdu intensivnostyu, prodolz-hitelnostyu i povtoryaemostyu livney,* Trudy NIU GUGMS, 1941.
2. Alexejev, G. A., *Metodika raschota maximalnykh razkhodov vody po krivym reduktsiyi osadkov,* Trudy GGI, Leningrad, 1963.
3. András, G., *Lefolyás és eróziv vizsgálata különbözo növények alatt és a talajfelszin különbözo allapota esetén,* Gedöllö, 1961.
4. Ayres, Q. C., *Soil Erosion and its Control,* New York-London, 1936.
5. Ayres, Q. C. and Coates, D., *Land Drainage and Reclamation,* McGraw-Hill, New York, 1939.
6. Bennet, H. H., *Soil Conservation,* New York-London, 1939.
7. Bisal, F., The Effect of Raindrop Size and Impact Velocity on Sand Splash, *Canadian Journal of Soil Science,* 40, 1960.
8. Blanchard, D. C., Behaviour of Water Drops at Terminal Velocity, *Transactions of the American Geophysical Union,* 31, 836, 1950.
9. Bratránek, A., Hydrological Data for Dimensioning River Training, *Technický obzor,* No.10, 1933.
10. Cablík, J. and Jůva, K., *Soil Conservation,* Prague, 1963.
11. Cormack, R. M. M., The Mechanical Protection of Arable Land, *Rhodesian Agricultural Journal,* 48, 1951.
12. Čížek, P. *Hydrology of Sewer Networks,* Prague 1961.
13. Dakshinamurti, C. and Biswas, T. D., Soil Erosion and Infiltration as a Function of Rainfall, *Proceedings of the Symposium on Land Erosion,* IASH, Bari, 1962.
14. Dyemidov, L. G. and Šigerin, G. G., *Kanalitsaciya,* Vol.I, Moscow, 1949.
15. Dub, O., *General Hydrology of Slovakia,* SVTL, Bratislava, 1955.
16. Dub, O., *Hydrology, Hydrography, Hydrometry,* Bratislava, 1963.
17. Dub, O. and Němec, J., *Hydrology,* TP, Praha, 1969.
18. Dvořák, J., *Research of Hydrological Data for Soil Conservation of Farm Plots,* Vědecké práce VÚM, Prague, 1962.
19. Dzubák, M., Research Problems in the Framework of the Czechoslovak National Program for IHD, *Hydrological Data for Water Resources Planning,* Praha, 1966.
20. Ellison, W. D., Studies of Raindrop Erosion, *Agricultural Engineering,* 1944.
21. Fair, G. M. and Geyer, L.Ch., *Water Supply and Waste-Water Disposal,* New York, 1956.
22. Förster, G., *Das forstliche Transportwesen,* Wien, 1885.
23. Gunn, R. and Kinzer, G. D., Terminal Velocity of Water Droplets in Stagnant Air, *Journal of Meteorology,* 6, 243, 1949.
24. Haeuser, J., *Kurze starke Regenfälle in Bayern,* München, 1919.

25. Härtel, O., *Die Wildbach u. Lawinenverbaung*, Tübingen, 1925.
26. Hellmann, G., *Die Niederschläge in den Norddeutschen Stromgebieten*, Berlin, 1906.
27. Hofmann, A., *La sistemazione idraulicoforestale dei Bacini Montani*, Torino, 1936.
28. Holý, M., *Water Erosion in Czechoslovakia*, MLVH, Praha, 1970.
29. Holý, M. and Vítková, H., Observations of the Effects of Slope on the Intensity of Erosion Processes, *International Symposium on Water Erosion*, ICID, Praha, 1970.
30. Holý, M., *Erosion Control*, SNTL/Alfa, Praha, 1978.
31. Hudson, N., A Review of Methods of Measuring Rainfall Characteristics related to Soil Erosion, *Research Bulletin* 1, Department of Conservation, Salisbury, Rhodesia, 1964.
32. Hudson, N., *Soil Conservation*, BT Batsford Ltd., London, 1973.
33. Komarov, V. D., *Gidrologicheskiy analiz i prognoz vesenovo polovodya ravninnych ryek*, Gidrometeoizdat, Leningrad, 1955.
34. Kostyakov, A. N., *Osnovy Melioratsiy*, Moskva, 1951.
35. Kozlík, V., *Economy and Effectiveness of Ditching in Erosion Control*, VÚV, Bratislava, 1959.
36. Krusekopf, H. H., *The Effect of Slope on Soil Erosion*, Columbia, 1943.
37. Kuron, H., Jung, L. and Schreiber, H., *Messungen von oberflächlichem Abfluss und Bodenabtrag auf verschiedenen Böden Deutschlands*, Hamburg, 1956.
38. Kuzmin, P. P., *Vodniye svoystva sněga*, Trud GGI 55/1956, Leningrad, 1955.
39. Kuzmin, P. P., *Intenzivnost snegotaniya v usloviyakh listvennogo lesa*, Trud GGI 55/1956, Leningrad.
40. Laws, J. O., Measurements of Fall-Velocity of Water-drops and Raindrops, *Transaction of the American Geophysical Union* 22, 1941, 709.
41. Linsley, R. K., Kohler, M. A. and Paulhus J. L. Ch., *Applied Hydrology*, McGraw Hill Book Comp., 1944.
42. Mezentsev, V. S., K teorii formirovaniya poverkhnostnogo stoka so sklonov, *Meteorologya i gidrologya* 3, 1969.
43. Mircchulava, C. E., *Inzheněrniye metody rascheta i prognoza vodnoy eroziyi*, Moskva, 1970.
44. Němec, J., *Hydrology*, SZN, Praha, 1965.
45. Neal, T. H., Effect of Degree of Slope and Rainfall Characteristics on Runoff and Soil Erosion, *Agricult.Eng.* 5/1938.
46. Philip, J. R., The Theory of Infiltration, *Soil Sci.*, USA, 1957.
47. Polyakov, B. V., *Gidrologicheskiy analiz i raschety*, Leningrad, 1946.
48. Polyakov, B. V., *Gidrologicheskiye raschety pri proyektirovaniyi sooruzheniy na rekách malykh basseynov*, Moskva, 1948.
49. Popov, E. G. and Velikanov, T. J., *Opyt priblizhennogo rascheta intenzivnosti snegotaniya v riechnom basseyně*, Trudy CIP, 22/49, Leningrad, 1950.
50. Pretl, J., Proposal for a New Method of Forecasting the Size of Soil Splash in Czechoslovak Conditions, Doctorship thesis, Civil Engineering Faculty, Czech Technical University, Prague, 1973.
51. Ramser, C. E., Runoff from Small Agricultural Area, *Journal of Agricultural Research* 34, 1927.
52. Schönfeldt, J., *Deutsche Wasserwirtschaft und Landwirtschaft*, 1934.
53. Skatula, L., *Damming Streams and Ravines*, SZN, Prague, 1960.
54. Slastichin, V. V., *Voprosy melioraciyi sklonov Moldavii*, Kišiněv, 1964.
55. Specht, A., *Die grössten Regenfälle in Bayern 1899 und 1901 und deren Benützung zur Berechnung der grössten Hochwassermengen*, 1915.
56. Sus, N. I., *Eroziya pochvy i borba s něyu*, Moskva, 1949.
57. Trupl, J., Intensity of Short Rains in the Elbe, Oder and Morava Catchments, *Práce a studie VÚV*, Praha, 1958.
58. Velikanov, A. M., *Gidrologiya suši*, Leningrad, 1964.
59. Wang, F., Die Ermittlung der Wasserabflussmengen, *Oesterreichische Vierteljahrsschrift für Forstwesen*, 1895.

60. Wischmeier, W. H., Punch Cards Records Runoff and Soil Loss Data, *Agricultural Engineering*, 26, 1955.
61. Wischmeier, W. H. and Smith, D. D., Rainfall Energy and its Relation to Soil Loss, *Transaction AGU*, 39, 1958.

3.2 MORPHOLOGICAL FACTORS

Water erosion is conditional on surface runoff from slopes. With increasing slope gradient and slope length and with continuing rainfall water running off the slopes gains higher velocity and tangential stress and the action of its destructive force on the soil surface increases. The intensity of erosion processes usually decreases with a drop of the slope gradient until soil particles which have been detached and transported over the soil surface begin to sediment. The course of erosion processes shows that areas most affected by water erosion have a rugged relief which enhances the concentration of surface runoff and accelerates runoff.

The morphology of the area also influences wind erosion whose intensity is affected by the exposure of the area to prevailing winds and the relief of the area.

3.2.1 *The Effect of Slope Gradient*

Theoretical studies and analyses of the effects of slope gradient on water erosion and numerous field observations and measurements as well as laboratory experiments have shown that slope gradient is one of the major erosion factors. Its effects on the initiation and course of erosion processes may be reduced by other factors, such as soil properties, the soil vegetative cover, etc. but never fully suppressed.

The interdependence of slope gradient and the erosion intensity as given by various authors shows that the intensity of the erosion process increases with growing tangential stress and velocity of the surface runoff which are prevalently the function of slope gradient. The importance of slope gradient for erosion intensity was proved by H. H. Bennet[1,2] in field measurements. The results of his measurements are given in Table 3.6.

TABLE 3.6 Effect of slope gradient on soil loss after
H. H. Bennet

Soil and location	Period of observation (years)	Rainfall (mm)	Length (m)	Slope (%)	Crop	Soil loss (t ha^{-1})
Silt loam — Muskingum, Ohio	9	965	22.1	8.0 12.0 20.0	maize	158.8 222.4 243.7
Fine sandy loam — Kirvin, Texas	10 8	1 032 1 092	22.1	8.7 16.5	cotton	50.1 136.8
Loam — Shelby, Missouri	14 10	1 025 749	27.6 22.1	3.7 8.0	maize	44.1 114.0

Similar conclusions have been arrived at by other authors, namely G. W. Musgrave[12], A. W. Zingg[20], T. H. Neal[13] and others who used field measurements and experiments on soil monoliths to derive the empirical relation between soil loss Sp and slope gradient I

$$S_p = f(I^n) \tag{3.26}$$

where n is within the region of 0.8-1.5.

W. H. Wischmeier and D. D. Smith[18] processed a large number of data on the intensity of erosion processes and on the basis of their studies they expressed the effect of slope gradient on soil loss by the equation:

$$S_p = f(\frac{0.43 + 0.30I + 0.043I^2}{6.613}) \tag{3.27}$$

B. V. Polyakov[14] conducted 111 measurements of soil loss on different slopes and arrived at the conclusion that erosion intensity is proportional to the squared slope gradient. The same was proved by I. S. Kostin[8] who measured this relation on artificial experimental plots 2-6 m^2 and in field conditions on experimental plots of 1 ha.

The decisive influence of slope gradient on the origination and course of erosion processes led to the determination of what is described as the critical slope, i.e., that slope on which a dangerous depletion of the soil surface occurs. It may be assumed that the dangerous depletion of the soil surface begins at that point at which surface sheet runoff changes into concentrated runoff, and sheet erosion changes into rill erosion. Data which have been processed show that the critical slope for acute erosion on low resistant soils ranges from 1 to $2°$, on medium resistant soils from 3 to $5°$ and on resistant soils from 6 to $7°$.

The given values are merely informative as they do not consider a number of factors which have a significant effect on soil erodibility.

3.2.2 The Effect of Slope Length

At constant slope gradient and under other constant conditions and providing rainfall duration is longer than the time needed for the water particles to travel from the water divide to the foot of the slope, the runoff, its intensity and tangential stress increases with slope length. This leads to an increase in erosion intensity. The relation between slope length and erosion intensity has been studied by many authors. A. S. Kozmenko[9] investigated the relation on plots with loamy-sandy soils and a slope of 5% (Fig. 48). The rapid break in the soil loss curve may be expressed by the development of sheet erosion into rill erosion effected by a greater concentration of surface runoff. Similar conclusions have been arrived at by A. S. Sobolev[16] and H. H. Bennet[2] (Table 3.7).

W. H. Wischmeier and D. D. Smith[18] made long-term field observations on the basis of which they derived the following expression for the relation between soil loss and slope length. The expression is written

$$S_p = f [(\frac{L}{22.13})^\alpha] \tag{3.28}$$

where L is the slope length measured from the water divide of the slope (m)

   α is the exponent dependent on the slope gradient; for $I \leq 10\%$ it is 0.5, for $I > 10\%$ it is 0.6.

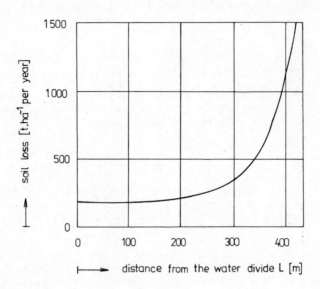

Fig. 48.   Relation between slope length and erosion intensity
           after A. S. Kozmenko.

TABLE 3.7 Effect of slope length on soil loss after H. H. Bennet

Soil and location	Period of observation (years)	Rainfall (mm)	Crop	Length (m)	Slope (%)	Soil loss (t ha^{-1})
Silt loam — Marshall, Iowa	1933-35	684	maize	48.0 96.0 192.0	8	28.9 40.3 52.6
Fine sandy loam— Vernon, Oklahoma	1931-36	800	cotton	11.0 22.1 44.3	7.7	42.5 55.6 95.3
Fine sandy loam— Kirvin, Texas	1931-36	1 038	cotton	11.0 22.1 44.3	8.7	45.9 68.9 107.7
Silt loam— Clinton, Wisconsin	1933-36	820	maize	11.0 22.1 44.3	16	159.0 248.0 286.6
Silt loam— Shelby, Montana	1934-35	851	maize	20.4 54.9 82.2	10	56.8 133.7 164.0

V. Horváth and B. Erödi[7] who investigated the effects of slope length on the amount of annual soil loss on plots of various lengths with a gradient of between 2-16% found that the amount of soil loss doubled with the quadruple increase of slope length.

The decrease in the amount of soil loss as related to the increase in slope length was observed by A. I. Spiridonov[17] and explained by N. I. Makayev[11] who observed that the most significant turbulence of precipitation runoff occurs in the upper part of the slope. This increases the saturation of runoff with detached soil particles which is bigger in the upper part of the slope than in the middle and lower parts where the soil surface is protected from the effect of raindrops by the water mantle. This view can only be accepted on the assumption that we are dealing with a continuous sheet surface runoff with increasing depth.

3.2.3 *Combined Effects of Slope Gradient and Slope Length*

The effects of the slope gradient and slope length on the intensity of the erosion process are significant for suggesting the type and location of erosion control measures. A number of expressions have been derived for calculating the combined effect of the two factors. One such expression by G. W. Musgrave[12] is written

$$S_p = k_p k_v I^{1.35} L^{0.35} Z_{30}^{1.75} \qquad\qquad (3.29)$$

where S_p is the erosion process intensity

$\quad k_p$ is the coefficient dependent on soil properties

$\quad k_v$ is the coefficient dependent on vegetative cover

$\quad I$ is slope gradient

$\quad L$ is slope length

$\quad Z_{30}$ is rainfall with a 30-min duration and maximum intensity
$\qquad$ occurring with a 2-year frequency.

A reliable investigation of the relationship between erosion process intensity, slope gradient and slope length and the knowledge of permissible soil loss allows the determination of this limit to be made for slope length. It is usually described as the critical slope length and its determination will be stated later in Chapter 9.

3.2.4 *Shape of Slope*

The intensity of erosion processes and its course is affected by the shapes of slopes, i.e. convex, concave, straight and combined (Fig. 49).

This classification allows us to observe the different course of erosion processes; this is because the prevalent erosion factor, i.e., the slope gradient, attains maximum values at different distances from the water divide depending on its shape (with the exception of straight slopes where it is constant) and it pertains to catchments of different sizes. The maximum effect of erosion processes will be manifest in those parts of the slope where the relation between the gradient and the distance from the water divide is the most propitious.

Convex slope A has in its upper part 1 a relatively small slope gradient and also a small distance from the water divide. As compared with other parts of the slope it is loaded with a small amount of runoff. Neither slope gradient nor slope length allow the full development of erosion processes. In the middle part of slope 2, slope gradient and slope length increase. Slope gradient and slope length reach

their highest values in the bottom part of the slope 3 where the intensity of
erosion processes reaches its maximum.

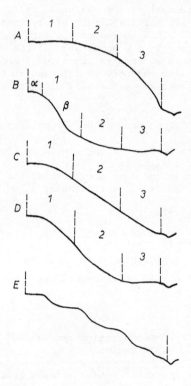

Fig. 49. Various shapes of slopes.

In concave slope B, the upper part of the slope may be divided into two sections.
The section at the watershed has a relatively small slope gradient which rapidly
increases and reaches its maximum in the lower section 1β. The gradient allows
the full development of erosion processes. Maximum slope length does not correspond
to maximum slope gradient as was the case with the convex slope. In the middle
part of slope 2 the slope becomes milder while slope length increases. The
intensity of erosion processes is determined by the relation between the decrease
in steepness and increase in slope length. In the lower part of slope 3, steepness
decreases to such an extent that material is deposited at the foot of the slope
even in the case of maximum slope length.

The straight slope C has an approximately constant slope gradient throughout.
Maximum erosion intensity may be expected at the point where the tangential stress
of surface runoff reaches its maximum.

Combined slopes have different shapes. Investigators should consider them separately. Most frequent are convex-concave slopes and graded slopes.

Convex-concave slope D has in its upper part 1 a relatively small gradient which increases in the middle part of the slope 2 where it reaches maximum. Here the convex shape develops into the concave shape and slope gradient decreases with slope length. Maximum erosion intensity may be expected in the middle part of slope 2.

In graded slope E, slope steepness alternately decreases and increases with growing slope length. The assumed course of the erosion process in the individual parts of the slope effect a continuous change in erosion intensity.

The said observations were confirmed by M. Hol\'y[5] by investigations made of the movement of soil particles (smaller than 0.01 mm) and nutrients (Fig. 50 and 51 a,b and 52) on slopes of different shapes in Czechoslovakia.

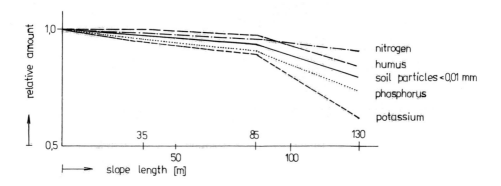

Fig. 50. Movement of soil particles and nutrients on convex slopes.

Comparisons made of erosion intensity on different slopes have shown that highest intensity is reached on convex slopes, lowest intensity has been observed on concave slopes at the same length and elevation.

The effect of shape on soil moisture and thereby on surface runoff and erosion intensity is significant. The difference between the convex and the concave slope cultivated in the same manner has been observed by Soviet authors([3 etc.]) (Tables 3.8 and 3.9).

The effect of the relief on wind erosion is given by the effect of the relief on the angle of the action of the wind current on the soil surface and on soil moisture which is determined by the depth and distribution of the snow cover on slopes of different shapes. On the concave slope the depth of the snow drift layer increases downslope. In the valley at the foot of the slope the depth of the snow layer is several times deeper than in the middle part of the slope. The more convex the slope the more the snow cover is blown away which reduces the depth of the snow layer downslope. On the convex-concave slopes wind blows the snow from the convex parts into the concave parts. This action of the wind is very significant because

the snow layer increases soil moisture and thereby also the consistence of the
soil and the growth of the vegetative cover which has positive effect on the erosion resistance of the soil.

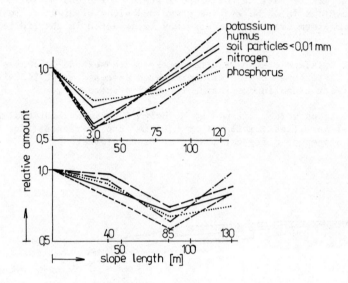

Fig. 51 a and b.   Movement of soil particles and nutrients
                   on concave slopes.

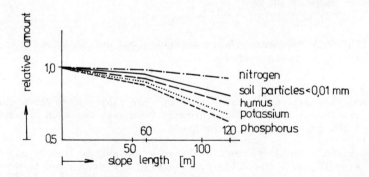

Fig. 52.   Movement of soil particles and nutrients on straight
           slope.

TABLE 3.8 Soil moisture on a concave slope

Slope section	Slope gradient (°)	Soil moisture (%)			
		in depth (cm)			
		10	30	50	average
Upper	4	6.8	13.5	16.3	12.8
Middle	1.5-2	7.3	14.8	17.2	13.7
Bottom	0	16.3	21.6	21.1	19.6

TABLE 3.9 Soil moisture on a convex slope

Slope section	Slope gradient (°)	Soil moisture (%)			
		in depth (cm)			
		10	30	50	average
Upper	4	13.7	19.4	23.0	18.8
Middle	5	12.4	18.0	20.3	16.8
Bottom	6	7.1	16.0	17.8	14.2

3.2.5 *Slope Exposure*

The exposure of the slope is given by its position to the north, south, east and west; slope exposure to solar radiation on the southern and western slopes causes the rapid thaw of snow resulting from differences in day and night temperatures. This results in higher surface runoff from the snow thaw, the freezing of vegetation and a more intensified depletion of the parental substrate which in turn increases the intensity of the erosion process as compared with shaded slopes turned to the north. Runoff from snow thaw is considerable especially from leeward slopes on which a deep layer of snow amasses over the winter. The soil of the sun-exposed slopes dries much quicker and this results in a more rapid decomposition of organic substances which in turn reduces its consistence and increases the danger of water and wind erosion.

Very important for the initiation and development of wind erosion is the exposure of the slope to the prevalent direction of the wind.

REFERENCES

1. Bennet, H. H., *Elements of Soil Conservation*, New York-Toronto-London, 1955.
2. Bennet, H. H., *Soil Conservation*, New York-London, 1939.
3. Gussak, V. B., Faktory i vnutrennyye posledstviya poverkhnostnych smyvov krasnozemev v usloviyakh vlazhnych subtropikov v Gruzii, *Sborník Eroziya počv*, AN SSSR, Moskva-Leningrad, 1937.
4. Holý, M., Effects of Slope Gradient on Water Erosion, *Vodní hospodářství*, 1,2,1955.
5. Holý, M., *Soil Conservation*, SNTL/ALFA, Praha, 1978.
6. Holý, M. and Vítková, H., Effects of Slope Gradient on Erosion Intensity, In: *International Symposium on Water Erosion*, ICID, Prague, 1970.

7.   Horváth, V. and Erödi, B., Determination of Natural Slope Category Limits by
     Function Identity of Erosional Intensity, Commission on Land Erosion IS, Bari,
     1962.
8.   Kostin, I. S., Charakteristika erozii pochv, *sbornik "Nauchnaya konferentsiya
     po izucheniyu i razvitiyu proizvoditelnykh sil Nizhnevo Povolzhya"*, Saratov,
     1945.
9.   Kozmenko, A. S., *Borba s eroziyey pochv*, Moskva, 1954.
10. Kuron, H., *Die Bodenerosion und ihre Bekämpfung in Deutschland*, 1941.
11. Makayev, N. I., *Ruslo reki i eroziya v yey baseyne*, Moskva, 1955.
12. Musgrave, G. W., The Quantitative Evaluation of Factors in Water Erosion, *Journ.
     of Soil and Water Conserv.* 2,3, 1947.
13. Neal, T. H., Effect and Degree of Slope and Rainfall Characteristics on Runoff
     and Soil Erosion, *Agricult. Eng.* 5, 1938.
14. Polyakov, B. V., *Gidrologicheskiy analiz i raschety*, Gidrometeoizdat, Leningrad,
     1964.
15. Silvestrov, S. I., *Eroziya i sevooboroty*, Moskva, 1949.
16. Sobolev, S. S., *Eroziya pochv v SSSR i borba s neyu*, Moskva, 1957.
17. Spirodonov, A. I., Opyt izucheniya vodnoy erozii i denudatsii v laboratorii,
     Pochvovedeniye 3, 1951.
18. Wischmeier, W. H. and Smith, D. D.,Predicting Rainfall-Erosion Loss from Crop-
     land East of the Rockey Mountains, *Agricult. Handbook* 282, Washington, 1965.
19. Zachar, D., *Soil Erosion*, SAV, Bratislava, 1970.
20. Zingg, A. W., Degree and Length of Slope as it Affects Soil Loss, *Agric.Eng.* 2,
     1940.

3.3 GEOLOGICAL AND SOIL FACTOR

The geological conditions of the area and the soil properties affect soil erodibility
and thereby also the intensity of erosion processes.

3.3.1 Geological Factor

The effect of geological conditions on the origination and course of erosion processes is manifest directly by the resistance of the denuded bedrock exposed to
the flow of water, air and indirectly affected by the character of the parent material whose properties are given by the bedrock.

Weathering bedrock often rises to the surface and is denuded by water and wind. In
such cases the surface is quickly disturbed and rills, gullies and ravines are
formed which spread and deepen very quickly.

The indirect effect of bedrock is manifest in the properties of the soilforming
parent material which conditions the principal properties of soils, namely their
structure, texture and the content of mineral and chemical substances which with
organic substances regulate the soil formation processes. The soils show varied
resistance to the action of surface runoff and wind erosion. Soils formed on
limestone and dolomite formations are relatively resistant; less resistant are soils
on igneous rocks and least resistant on various sediments, namely sandstone, loam,
clay and chalk, flysch formations and loess sediments.

Any evaluation of geological conditions on the origination and development of erosion processes must consider the formation of soils, as some genetic factors may
significantly reduce the influence of the parent material on soil properties. This
mainly occurs in areas with climatogenic soils where the significant influence of
the soilforming substrate on soil erodibility is only manifest in soils formed on
loess and in the flysch zone while its influence has not been proved on other geological formations[4]. The effects should therefore be investigated of geological

factors on water erosion in specific local conditions. The important role of geological conditions in the complex process of erosion is evident and should not be
ignored.

3.3.2 Soil Factor

Soil conditions which are the sum of the individual properties of the soil affect
the infiltration of precipitation into the soil and the resistance of soil to the
destructive effects of raindrops, surface runoff and wind.

The major factors affecting the infiltration of precipitation into the soil are
soil texture, structure, soil moisture and stratification. Next to these factors
the most important factors for erosion resistance are the humus content and the
saturation of the sorption complex.

Investigations on the effects of soil texture on erosion processes have shown that
sandy soils are least susceptible to erosion. This is because as compared with
other soils they are highly permeable; at low consistence the major proportion of
heavier soil particles resist the kinetic energy of water and the kinetic energy of
the wind for the longest period of time. On the other hand, clay soils with a low
permeability have a high content of colloidal particles and show a high level of
consistence in a mildly humid state. Loamy soils have medium permeability and a
considerably lower level of consistence resulting from a high content of silt particles. The least favourable properties with regard to soil erosion have been found
in loess with a lack of cementing colloidal particles. Their dispersion rate is
increased by leaching which results in the loss of the last bond, i.e., $CaCO_3$ and
humus.

The effect of the presence of particles of various sizes in the soil on its resistance to water erosion is considered by some authors as being so important that
they describe it as erodibility, i.e., the soil's vulnerability to erosion. Thus,
for instance, A. A. Cherkasov[3] recommends erodibility E_p to be expressed by the
erosion index written

$$E_p = \frac{d'h'}{a} \tag{3.30}$$

where d' is the ratio of the fraction of soil particles with grain size < 0.05 mm
        obtained by analysis without treating the sample with NaCl to the same
        fraction obtained after treatment with NaCl
      h' is the numerical expression of water absorption rate of soil per 1 g of
        colloidal particles contained in it, i.e., the so-called index of soil
        hydrophilic capacity
      a is the amount of soil aggregates resistant to the action of water in a
        unit of soil.

J. G. Bouyoucos[2] expresses soil erodibility by the ratio of sand particles (0.06–
2.0 mm) and loam particles (0.002–0.06 mm) to the content of clay particles (<0.002
mm).

American soil conservation experts (cit.6) usually determine the soil erodibility
by the expression

$$ER = DR : CE \tag{3.31}$$

where ER is the erosion ratio

      DR is the dispersion ratio

CE is the moisture equivalent.

Soil erodibility increases with ER and the boundary between soils susceptible to water erosion and resistant soils is approximately ER=10.

The study of the effects of the soil on water erosion should include an investigation of the entire soil profile. In areas where the soil consists of layers with different textures the water erosion resistance of the investigated locality is affected by the stratification of the layers, and in case of a shallow soil relief also by the properties of the parent material. Where a permeable layer rests on an impermeable layer and where there is a considerable amount of infiltration, the surface layer may become oversaturated with water which the lower layer is unable to absorb. This leads to an intensive wash of the permeable layer. This most frequently occurs in podzolic soils where the more permeable A horizon rests on the compact B horizon.

The evaluation of the effects of soil texture on the formation of erosion processes should include an assessment of the effects of the soil morphology. Skeletal soils which contain a low proportion of colloidal fractions are usually characterized by high permeability and low cohesion. Soil permeability and reduced soil particle mobility often results in a lower intensity of erosion processes.

The soil structure which is determined by the arrangement of soil particles into aggregates determines the content of noncapillary pores in the soil and the stability of the soil aggregates. Soils with a favourably developed structure retain more water and are more resistant to the destructive force of surface runoff and to wind than soils with an insufficiently developed structure.

Experience and numerous measurements have proved that the favourable effects of soil structure are most explicitly manifest in soils with a crumb structure where precipitation infiltrates through non-capillary pores to the deeper soil layers while a considerable proportion of rainfall is retained in the capillary pores of the crumbs which in turn secures favourable soil moisture and thereby also soil cohesion. V. R. Vilyams[12] says that soil with a crumb structure retains up to 85% of rainfall. L. Smolik[10] found that soils with a crumb structure have full infiltration capacity and erosion resistance at a 17% slope gradient. Loess soils have very low resistance to wind erosion. V. Novák[9] found that at a velocity of 5 m s^{-1} the wind blows soil particles with a grain size of 0.25 mm, at a velocity of 9 m s^{-1} particles with a grain size of 0.75 mm and at a velocity of 12 m s^{-1} particles with a grain size of 1.5 mm.

The formation and preservation of a crumb soil structure is conditional on favourable physical, chemical and biological soil properties. These are affected by the presence of mineral and colloidal substances, namely the content of effective humus. It is therefore useful to conduct next to analyses of the soil aggregates, the analyses of the content of effective, i.e. calcium saturated humus which holds together the soil particles and is the basic prerequisite for their resistance and of exchange calcium which results in the coagulation of soil colloids and thereby in the formation of solid aggregates.

Many authors have tried to use soil structure as the basic criterion for explaining soil erodibility. D. G. Vilenskiy[11] recommends either the investigation of the erosion resistance of the individual aggregates or the investigation of the rate of the breaking down of aggregates in samples of undisturbed soil exposed to the effects of simulated rainfall. V. B. Gussak (cit. 6) measured the quantity and velocity of water needed for the detachment and removal of the given amount of soil in a specially designed flume, J. K. Alderman[1] studied the size of a crater formed by a jet of water in a soil sample, etc.

Soil texture and structure are characteristic for soil types and may become important indicators of the erosion susceptibility of the soil. Investigations have
shown that under the same conditions chernozem soils which have a crumb structure
and considerable cohesion are most resistant to erosion while brown soils which
usually do not have a highly developed soil structure and have lower cohesion show
lower erosion resistance, and podzolic soils with a loess structure and incohesion
of soil particles show least erosion resistance.

The intensity of water and wind erosion depends on soil moisture which affects the
value of the runoff coefficient and significantly affects soil cohesion. Excessively
high soil moisture reduces the infiltration of rainfall into the soil thereby
increasing the surface runoff and breaking down soil aggregates. Low soil moisture
reduces the resistance of soil, namely to wind erosion. Investigations have shown
that soil aggregates show maximum resistance to erosion at such soil moisture levels
at which optimal soil structure is formed.

REFERENCES

1. Alderman, J. K., The Erodibility of Granular Material, *Journal of Agr.Eng.*, 1,
    1956.
2. Bouyoucos, J.G., The Clay Ratio as a Criterium of Susceptibility of Soils to
    Erosion, *Journal of Am.Soc. of Agronomy*, 27, 1935.
3. Cherkasov, A. A., *Melioratsiya i selskochozyaystvenoye vodosnabzhnie*, Moskva,
    1950.
4. Holý, M., Investigation of Soil Erosion in Northern Bohemia, *Annual Report on
    Research*, Task No.011638, Prague, 1954.
5. Holý, M., Hydrology of Erosion Phenomena, Doctorship thesis, Prague, 1964.
6. Hudson, N., *Soil Conservation*, London, 1973.
7. Kuron, H., *Die Bodenerosion und ihre Bekämpfung in Deutschland*, Kulturtechnik,
    1941.
8. Kutílek, M., *Water Conservancy Soil Science*, SNTL/SVTL, Prague, 1978.
9. Novák, V., *Soil Science*, Text-book, Brno, 1951.
10. Smolík, L., Contribution to a Better Knowledge of Structural Soil Units, *Collec-
    tion of Works*, ČSAV, Prague, 1948.
11. Vilenskiy, D. G., *Pochvovyedyeniye*, Moskva, 1950.
12. Vilyams, V. R., *Osnovy zemledeliya*, Moskva, 1950.
13. Zachar, D., *Soil Erosion*, Bratislava, 1970.

3.4 VEGETATIVE COVER

The vegetative cover protects the soil surface from the direct impact of raindrops
and from the effects of wind. It enhances the infiltration of rainfall into the
soil and slows down surface runoff and thereby improves the physical, chemical
and biological properties of the soil. Very important is the soil binding effect
of the root system of the vegetation. In the winter season the vegetative cover
effects an even distribution of the snow cover and reduces the danger of soil
freezing.

The soil surface is protected from the impact of raindrops by the interception of
the above-ground organs of plants (Fig. 53). The plant bodies damp the energy of
raindrops, thereby reducing the danger of the breaking down of soil aggregates.
The raindrops are caught by the plants and from there slowly slide to the soil
surface. This protraction together with the slowing down of the velocity of
surface runoff owing to the increased roughness of the soil surface by the above-
ground plant organs, supports the infiltration of water into the soil and as a
result reduces surface runoff. Increased infiltration results from soil properties

improved by the vegetative cover. This is because the vegetation enriches the soil with organic substances and nitrogen and causes the movement of substances (e.g., CaCO₃) which contain favourable properties, from the deeper ineffective soil layers to the soil surface. It aerates the soil, induces increased microbial activity, etc., all of which influence the formation of the soil structure and improve soil cohesion.

Fig. 53.   Interception of water by above ground organs of
           plants (photo by courtesy of the Soil Conservation
           Service, USA).

By shading the soil the vegetative cover reduces evaporation and conserves moisture which significantly affects the stability of soil aggregates.

The favourable effects of the vegetative cover include the mechanical reinforcement of the soil by the root system. The important factor in the binding effect of the roots is the density of the root system and the depth of its penetration into the soil profile.

Vegetation also protects the soil from the effects of wind erosion by becoming a buffer between the soil and the wind. By improving the soil properties it increases its resistance to wind erosion.

The favourable effects of vegetation on the course and intensity of the erosion process differ according to type of vegetation and its condition. Field observations and measurements have made it possible to rank plant cultures by their erosion resistance effectiveness. Ranking highest is the forest, followed by grass, corn and potatoes.

The forest with a dense canopy, good undergrowth and undisturbed litter have the most significant effect on surface runoff and thereby on the intensity and course of erosion[6]. Surface runoff from forest soil does not, as a rule, exceed 10% of total precipitation. J. Cablík and K. Jůva[3] who have made this observation have thus confirmed that forest soils with a good forest cover do not suffer from water erosion. C. R. Hursch (cit. 9) found that when the experimental plot was stripped of forest cover and replaced by pasture, maximum surface runoff from the plot increased over 7 years from 0.33 to 20.0 m^3s^{-1} km^2, soil loss increased 24-fold and after intensive rainfall to 500-fold.

Forest growths have significant effect on surface runoff from the catchment where by increasing infiltration of rainfall and by its interception and retarding surface runoff they reduce maximum runoff. This results in the decreased intensity of erosion processes in the catchment area and especially in the river beds. This favourable effect of the forest growth has been proved in experimental catchments in Czechoslovakia[8]. The erosion effects of runoff from the catchments of two streams having approximately the same area (400 ha), the same conditions except for forest cover — in the one forest growth covers 93.2% of the area, in the other it only covers 4.7%— differed considerably and proved the favourable effects of forest growth on runoff and thereby on the intensity of erosion processes (Fig. 54 a and b).

Similar observations on the protective effects of forest growths were made in other catchments where the inter-actions were monitored between meteorological factors and runoff on slopes[10].

The effects of grass cover with a well-developed turf on the value and course of surface runoff and of soil conservation are similar to those of forest growths. H. H. Bennet[2] found that surface runoff from plots with a good grass cover ranged from 0.3 to 5.5% of total precipitation and soil loss amounted to 0.029-0.132 t ha^{-1} while under the same conditions runoff from forest growths was 0.1-3.6% and soil wash was 0.005-0.193 t ha^{-1}. There was no marked difference between the efficiency of the forest and of the grass cover.

In a model area M. Holý and J. Váška[4] conducted a comparative investigation of runoffs from a bare plot and from a plot with grass cover. They found that grass cover has a significant effect on surface runoff. In the experimental period (1960-1969) runoff from the plot under grass (44.5% gradient, area 20 x 6 m) was 96% less than runoff from the plot (same gradient, same area) bare of vegetation.

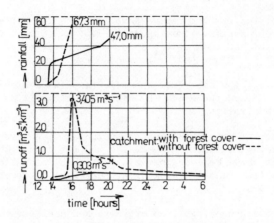

Fig. 54 a and b. Effect of forest growth on runoff.

The poor protective cover given to the soil by field cultures is due to their small vegetative canopy per unit of soil surface, the usually short above-ground parts of the plants in most seasons (these are usually annual plants) and sparse thin roots. The erosion control effect of field plants by the size of their leaf surface per 1 m^2 of soil surface as observed by Werner (cit. 2) is given in Table 3.10.

TABLE 3.10   Vegetative conopy per 1 m^2 of soil surface after Werner

Beet	1.6 m^2	Rye	15.6 m^2
Rape	1.7 m^2	White clover	19.6 m^2
Maize	11.7 m^2	Meadow clover	26.4 m^2
Barley	14.4 m^2	Alfa-alfa	85.6 m^2

All results obtained from investigations of the vegetative cover and its impact on erosion intensity conducted under different conditions show that vegetation has a favourable effect on erosion control. These data also proved the accuracy of the following order of cultures with regard to their erosion control effects: forest growths, permanent grass, grass planted in rotation with other plants, corn, potatoes.

Investigations of the importance of vegetation for erosion control in concrete conditions must be considered with regard to the type and condition of the vege-tation in the season during which the soil is most affected by erosion.

REFERENCES

1.   Arman, P., *Unde masuri de combatars a erosiunii solului in cimpia Transilvanici*, Bucharest 1956.

2. H. H. Bennet, *Elements of Soil Conservation*, New York-Toronto-London, 1955.
3. Cablík, J. and Jůva, K., *Soil Erosion Control*, Prague, 1963.
4. Holý, M. and Váška, J., Relations between Surface Runoff and Soil Loss in
     Water Erosion, *ICID Water Erosion Symposium*, Prague, 1970.
5. Kuron, H. and Jung, L., *Über die Erodierbarkeit einiger Böden*, Comptes Rendus,
     Toronto, 1957.
6. Mráček, Z. and Krečmer, V., *Importance of the Forest for Human Society*, SZN,
     Prague, 1975.
7. Musgrave, G. W., The Quantitative Evaluation of Factors in Water Erosion,
     Journ. of Soil and Water Conservation 2,3, 1947.
8. Válek, Zd., An Investigation of the Impact of the Forest on Runoff in the
     Kýchová and Zděchovka Catchments, *Vodní hospodářství* 10-11, 1953.
9. Zachar, D., *Soil Erosion*, Bratislava, 1970.
10. Zelený, V., *The Impact of the Forest on Runoff and Erosion in the Beskydy*,
     Prague, 1964.

3.5   TECHNICAL FACTOR

Technical conditions consist mainly in the methods of land-use and land tillage,
the choice and distribution of cultures and the agrotechnology. Technical conditions are a typical anthropogenic factor which may positively or negatively affect
erosion intensity.

Erosion is most intensive on soil on which the former growth has been disturbed,
i.e., mostly on agricultural soils, on soils stripped of growth for other reasons,
e.g., housing construction, communications, training fields, etc. Any interference
in the natural vegetative cover should be considered with regard to its possible
consequences which usually result in intensive erosion, and any future project
which envisages such a disturbance of the vegetative cover should provide for
erosion control measures.

Intensive wind and water erosion mostly affecting large areas usually results from
the conversion of countryside with a natural vegetative cover, mostly forests, into
farmland which is then intensively tilled. Disasters caused by the deforestation
of the Balkans, the conversion of wild virgin countryside in USA into corn monocultures, dust storms in the deserts, semi-deserts and steppes of the whole world
are well-known[1].

The intensity and course of erosion processes in agricultural soil is significantly
affected by the position and shape of the plot. Observations have shown that fields
on slopes whose longer side is along the contour and which are contour farmed suffer
considerably less from erosion than plots tilled down slope in straight lines.
Erosion intensity in contour farming is much less than in slope tillage. A. A.Cherkasov
[3] arrived at conclusions given in Table 3.11.

TABLE 3.11   Effect of cropping method on soil loss after
             A. A. Cherkasov

Location of rows	Soil loss (t.ha^{-1})
Along contour	1.7
On slope with 4.4% gradient	12.2
On slope with 16% gradient	27.2

The effects of contour farming are given by the higher retention capacity of the soil surface and its higher roughness. Surface runoff has to overcome obstacles in form of ditches, rows, etc., which retard surface runoff and increase the infiltration of rainfall into the soil which in turn affects soil moisture and the stability of aggregates.

The erosion resistance of the soil may be increased by purposefull treatment.

Erosion intensity is also affected by the location of the farm culture. Water erosion is less when erosion resistant cultures are planted on slopes which are most susceptible to erosion, wind erosion may be reduced by planting wind erosion resistant crops on sites which are likely to suffer most from wind erosion. Seeding procedures should not ignore these facts.

Any assessment of the technical factor must consider the efficiency of erosion control measures which are being implemented. The efficiency of these methods is represented by the relation of erosion intensity on a plot protected by the given soil conservation measures to soil erosion on a plot without conservation measures having the same soil properties and treated downslope. O. Stehlík and M. Šabata have observed[5] that if the value of the coefficient of erosion control measures in downslope tillage is considered as being equal to one then contouring provides 50% erosion control, ditching 65%, strip cropping 75%, terracing (without retention area) 80%, terracing (with retention area) up to 90%.

The effect of conservating measures on erosion intensity differs.

Correctly applied land improvement, such as drainage, irrigation and the reclamation improves runoff conditions and soil properties, thereby reducing erosion intensity.

Communications with a steep gradient and unsuitable cross profile represent a great erosion hazard. Unless such communications are provided with appropriately reinforced embankments and with ditches to divert surface runoff then this surface runoff will be retained in the ditches which in time will turn into rills and gullies (Fig. 55).

On construction sites the soil is stripped of its vegetative cover, the earth is piled up into steep slopes, the soil surface is cut up, etc., all of which will result in a change of runoff conditions causing intensive erosion and often landslides. Such cases have of late become frequent on the building sites of highways and housing estates.

Project designers should carefully consider any interference in the natural environment if this proves necessary and should include erosion control measures in project design.

REFERENCES

1. Beasley, R. O., *Erosion and Sediment Pollution Control*, Iowa, USA, 1972.
2. Burda, F., Jukl, A., Kratochvíl, V., Škaloud, J., Uhlíř, P. and Vondřejc, J., Agricultural Production, *Zemědělská výroba*, SZN, Prague, 1960.
3. Cherkasov, A. A., *Melioratsiya i selskokhozyaystvennoye vodosnabzheniye*, Moskva, 1950.
4. Economic Commission for Europe, *Proceedings, Seminar on the Pollution of Waters by Agriculture and Forestry*, Vienna, 1973.
5. Stehlík, O. and Šabata, M., Factors Affecting Erosion, In: *Proceedings of Symposium on Erosion Control*, Brno, 1971.

Fig. 55. Gully started from a road (photo by courtesy of the
Soil Conservation Service, USA).

3.6 SOCIO-ECONOMIC FACTOR

The use and exploitation of natural resources is determined by the level of the
society and its system. The most effective exploitation of natural resources
requires that any interference in the natural environment should be in harmony with
the needs of society and that those who undertake it should have a profound knowledge of the laws of nature.

In favourable socio-economic conditions with a high standard of education, good
relations may be established between society and the basic natural resources, i.e.,

water and soil. In such a society it is also possible to implement an agricultural
planning and capital construction policy in all branches of the national economy
such as would ensure that the induced changes in the natural environment should
have minimal negative impact on human society owing to the depletion of natural
resources, namely soil erosion and water pollution.

4. Theory of Water Erosion

Erosion is caused by surface runoff and results from complex natural processes. The determination of its course, its mathematical representation and the prediction of erosion processes of a certain intensity and frequency in the given conditions is a complex hydrological problem. The process of water erosion results and develops owing to surface runoff which is affected by a number of factors and their interrelation. The theory of water erosion should therefore be oriented to the laws of sheet and concentrated runoff and to transport processes caused by water.

4.1 SHEET SURFACE RUNOFF

Surface runoff from slopes starts as sheet runoff and owing to the uneven surface of the soil surface develops into concentrated runoff.

Researchers investigating sheet surface runoff usually assume that the depth of water running off the slope at a constant gradient increases with the distance from the water divide (Fig. 56).

Velocity and tangential stress are important characteristics of sheet surface runoff.

The velocity of surface runoff may be expressed using Chezy's equation written

$$v_x = c\sqrt{RI} \qquad (4.1)$$

where v_x is the velocity of surface runoff at distance x from the water divide $(m\ s^{-1})$

    c is velocity coefficient $(m^{1/2}\ s^{-1})$

    R is hydraulic radius (m)

    I is slope gradient

Hydraulic radius R at small water depth y, flow cross-section P' and wetted perimeter 0, may be expressed for the surface of a unit width using the equation

$$R = \frac{P'}{0} = \frac{1 \cdot y}{1+2y} \doteq y \qquad (4.2)$$

ignoring small member 2y.

83

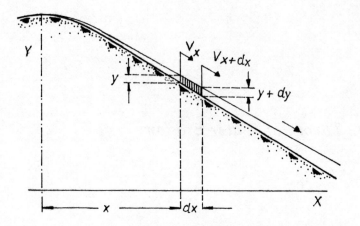

Fig. 56.   Diagram of sheet surface runoff.

The coefficient of velocity after H. Bazin may be written

$$C = \frac{87\sqrt{y}}{\gamma+\sqrt{y}} \doteq \frac{87}{\gamma}\ \sqrt{y} = m'\sqrt{y} \qquad\qquad\qquad (4.3)$$

ignoring the member $\sqrt{y}$ in the denominator and assuming that $\dfrac{87}{\gamma} = m'$.   The value of
the roughness coefficient γ for different soil surface are given by A. A. Cherkasov
(Table 4.1, cit. 7).

TABLE 4.1   Values of roughness coefficient γ after A. A. Cherkasov

State of soil surface	Coefficient γ	$m' = \dfrac{87}{\gamma}$
Fields tilled downslope	2.0	43.50
Tilled fields with levelled surface	3.5	24.85
Fields overgrown with reeds	4.0	21.75
Fields overgrown with moss	5.0–6.0	17.40–14.50
Meadows with low grass cover	6.0–8.0	14.50–10.88
Rough soil surface with molehills	8.0–15.0	10.88–5.80

Equation (4.1) may be expressed following the calculation of equations (4.2) and
(4.3).   The equation may be written as follows:

$$v_x = m'\sqrt{y} \cdot \sqrt{y}I = m'\sqrt{I} \cdot y = \propto y \qquad\qquad\qquad (4.4)$$

substituting m'. $\sqrt{I} = \propto$ at constant slope gradient and slope surface.

Using this method it is possible to obtain the relation between the velocity of
surface runoff and the water depth in any profile at distance x from the water
divide.   For erosion control it is useful to investigate the relation between velocity v_x and the distance from the water divide x.

The increment of precipitation runoff dQ in the elementary section dx (Fig. 56) is

$$dQ = (y + dy) \, v_{x+dx} - y v_x \qquad (4.5)$$

Having substituted velocity from expression (4.4) the following equation applies:

$$dQ = (y + dy)\alpha(y + dy) - y\alpha y = 2\alpha y \, dy + \alpha(dy)^2 = 2\alpha y \, dy \qquad (4.6)$$

ignoring the small variable of the second order.

The runoff quantity in section dx may also be expressed as the precipitation quantity in section dx reduced by infiltration and written

$$dQ = oi \, dx \qquad (4.7)$$

where i is rainfall intensity

   o is runoff coefficient.

From the quality of expressions (4.6) and (4.7) ensues

$$2\alpha y \, dy = oi \, dx \qquad (4.8)$$

following integration

$$y = \sqrt{oi \frac{x}{\alpha}} \qquad (4.9)$$

By substituting expression (4.9) into expression (4.4) we obtain the relation between the velocity of surface runoff v_x and distance x from the water divide which may be written

$$v_x = \sqrt{m' oix I^{1/2}} \; [\,\mathrm{m.s^{-1}}\,] \qquad (4.10)$$

Tangential stress of surface runoff may be expressed by the equation

$$\tau = \gamma_v y I \; [\,\mathrm{Nm^{-2}}\,] \qquad (4.11)$$

where γ_v is the specific gravity of water $(\mathrm{N \, m^{-3}})$

   y is the depth of surface runoff (m)

   I is the slope gradient (%).

By substituting equation (4.11) for y in expression (4.9) we obtain

$$\tau = \gamma_v I \sqrt{o \frac{x}{\alpha}} = \gamma_v I \sqrt{oi \; \frac{x}{m' I^{1/2}}} \; [\,\mathrm{Nm^{-2}}\,] \qquad (4.12)$$

which expresses the relation between tangential stress τ and distance from the water divide x.

4.2   CONCENTRATED SURFACE RUNOFF

Water running off the soil surface gradually concentrates owing to soil surface roughness until surface runoff becomes a stream and then a water course. Important for erosion control is surface runoff in stream beds which moves considerable quantities of sediments which are transported by the stream to lower altitudes.

Sediments are put in motion at a certain friction velocity of the stream water. This velocity depends mainly on the nature and size of the sediments and on their friction coefficient.

At extreme velocity of water an equilibrium is established between the force of the water flow on the sediments and the resistance force of the sediments. This relation is shown in Fig. 57 and may be written

$$G \sin\alpha = k' P_s \gamma_v \frac{v_{kr}^2}{2g} \tag{4.13}$$

where G is the gravity of sediments in dimensions a.b.c. (N)

α is the angle of the stream bed slope

k' is the coefficient of the shape of the sediments (for prism k' = 1.5, for circular ellipsoid k' = 0.80)

P_s is the impact surface of sediments
P = a.c (m^2)

γ_v is the specific gravity of water (N m^{-3})

g is acceleration due to gravity (m s^{-2})

v_{kr} is the extreme velocity (m s^{-1}).

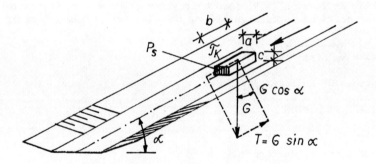

Fig. 57.   Forces affecting sediment transport.

From the equation (4.13) follows the expression which may be written

$$v_{kr} = \left(\frac{G \sin\alpha \; 2g}{k' \gamma_v \; P_s} \right)^{1/2} \tag{4.14}$$

from which after substituting G = V($\gamma'-\gamma_v$)

where V is the sediment volume (m^3)

γ' is the specific gravity of the sediment (N m^{-3}).

results in an equation which may be written

$$v_{kr} = \left(\frac{V(\gamma'-\gamma_v)f\cos\alpha}{k'\gamma_v} \frac{2g}{P_s}\right)^{1/2}$$ (4.15)

where f is the coefficient of the friction of sediments on the stream bed; f = 0.5 to 0.8.

The expression (4.15) for the given conditions may be written

$$v_{kr} = \xi \sqrt{b}$$ (4.16)

where coefficient ξ is dependent on the shape, gravity and friction of sediments.

For grains of a spherical shape with diameter r

$$v_{kr} = 4.46 \sqrt{r},$$

for grains of circular ellipsoid shape with the major semi-axis b

$$v_{kr} = 4.43 \sqrt{b},$$

for grains of a prismatic shape with the biggest dimension being ℓ

$$v_{kr} = 3.45 \sqrt{\ell}.$$

The relation between velocity v_{kr} and mean profile velocity v_s may be written v_{kr} = 0.50 to 0.75 v_s. Many authors have attempted to determine the values of critical velocity v_{kr}. The values of this velocity are derived from direct observations and are given in Table 4.2.

TABLE 4.2.   Values of extreme velocities

Type of sediment	$v_{kr}(\text{m.s}^{-1})$
Sand with grains the size of aniseed	0.108
Sand with grains the size of peas	0.189
Sand with grains the size of beans	0.325
Round boulders 0.027 m in size	0.650
Sharpedged boulders the size of eggs	0.975
Rough fist sized silt	1.400

For nonhomogeneous sediments which occur most frequently in stream beds I. I. Levi [23] gives the following relation:

$$v_{kr} \geq 1.4 \sqrt{gd_s} \lg \frac{h}{7d_s} \left(\frac{d_{max}}{d_{min}}\right)^{1/7} \, [\text{m.s}^{-1}]$$ (4.17)

where d_s is the mean diameter of the sediment (m)

d_{max} is the largest diameter of the sediment (m)

d_{min} is the smallest diameter of the sediment (m)

g   is acceleration by gravity (m s^{-2})

h   is the water depth (m).

From velocity v_{kr} it is possible to derive the so-called compensated slope gradient of the stream bed along which the sediments still move and do not yet settle but which is not receptive to any new sediment.

In the case of a compensated gradient the mean flow rate of water in the stream v_s equals the velocity v_{kr} of sediments forming the stream bed. This applies for the application of the Chézy equation and for the expression (4.15).

$$c\sqrt{RI} = \left(\frac{V(\gamma'-\gamma_v)\ f\cos\alpha\ 2g}{k'\gamma_v\ P_s} \right)^{1/2} \tag{4.18}$$

$$I_v = \frac{V(\gamma'-\gamma_v)\ 2g\ f\ \cos\alpha}{k'c^2\ R\gamma_v} \tag{4.19}$$

C. D. Valentini[37] gives the relation written

$$I_v = C'\ \frac{b}{R} \tag{4.20}$$

where I_v is the compensated gradient of the stream bed

b   is the edge of the mean cuboid boulder whose volume $b^3 = \frac{V}{n}$ is obtained as the share of the total volume V without any selection of the collected boulders and their number n(m)

C' is the value which Valentini obtained by measurement; to the 25% bed slope it is C' = 0.093

R   is the hydraulic radius (m).

The mean flow rate decreases with the increasing amount of sediments up to the saturation of the water with sediments.

The relation applies

$$v_s = v_s'\ \frac{\gamma_v}{\gamma_v + n(\gamma'-\gamma_v)} \tag{4.21}$$

where v_s is the mean flow rate following saturation with sediments (m s^{-1})

v_s' is the mean flow rate before loading by sediments (m s^{-1})

n   is the ratio of sediment volume to the water volume.

The so-called erodibility factor is derived from relation (4.21) and may be written

$$k_b = \frac{\gamma_v}{\gamma_v + n(\gamma'-\gamma_v)} \tag{4.22}$$

which should be applied for correcting the mean flow rate in sediment saturated flows.

The compensated slope gradient may also be calculated from the tangential stress of flowing water. To each type, shape and size of sediments pertains an extreme tangential stress at which equilibrium still prevails between the tangential stress and the gravity of sediments settled on the stream bed. The relation applies

$$\gamma_v RI = G\ \sin\alpha \tag{4.23}$$

where G is the gravity of sediments per area (N m^{-2})

$$G = V_n(\gamma'-\gamma_v)$$

sinα ≐ tgα =f

V_n is the amount of sediments per area.

From the relation (4.23) follows relation (4.24) which may be written

$$I_v = \frac{V_n(\gamma'-\gamma_v)f}{\gamma_v R} \qquad (4.24)$$

The values of critical tangential stress for certain types of material are given in Table 4.3.

TABLE 4.3. Values of extreme tangential stress for different materials

Materials	Grain size (mm)	Tangential stress (N m^{-2})
Fine grained siliceous sand	0.4-1.0	2.45-2.94
Course grained siliceous sand	to 2	3.92
Loam	-	9.81-11.77
Siliceous gravel	5-15	12.26
Turf planted for a short period of time	-	19.61-29.42
Willow cover	-	39.23
Course grained grit	40-50	47.07
Flat calcareous silt 1-2 cm in thickness and 4-6 cm in length	-	54.92
Stone pavement	-	98.07 and more

The conditions for the movement of sediments are depicted very accurately by E. Mayer-Peter and R. Müller[24] who give the relation between the hydraulic values of the flow profile and the transported amount of sediments G_s per 1m of bed width in kg m^{-1}s^{-1} at mean grain diameter d_m. The relation may be written

$$\gamma_v \frac{R_s I_r}{\gamma_s'' d_m} = \gamma_v \frac{Q_s}{Q} \left(\frac{k_s}{k_r}\right)^{3/2} \frac{h I_r}{\gamma_s'' d_m} = A''+B''\left(\frac{\gamma_v}{g}\right)^{1/3} \frac{G_s^{2/3}}{\gamma_s'' d_m} \qquad (4.25)$$

where γ_v is the specific weight of water (kg m^{-3})

R_s is the hydraulic radius which applies for sediment movement per 1m of bed width;

$R_s = h \dfrac{Q_s}{Q}$ (m)

h is the water depth (m)

I_r is the gradient (‰)

Q is the total flow rate (m^3 s^{-1})

Q_s is the flow rate affecting the movement of sediments, $Q_s = q_s B$ (m^3 s^{-1})

q_s is specific flow rate, $q_s = \dfrac{Q}{2h+B}$ (m^3 m^{-1} s^{-1})

B is bed width (m)

k_s is $\dfrac{21.1}{\sqrt{d_m}}$ after Wittmann; during flow of sediments coefficient k_s is variable, the value 21.1 drops down to 15

k_r is $\dfrac{26}{\sqrt{d_{90}}}$ (after E. Mayer-Peter)

d_m is the standard sediment average;

$$d_m = \frac{\Sigma(d_i \Delta p_i)}{100} \ [m]$$

d_i is the average of extreme values of sediment fractions (m)

Δp_i is the share of sediment fractions with diameter d_i (%)

d_{90} is the diameter for sediment for p_i = 90 % (m)

$\gamma_s" = (\gamma_s - \gamma_v)$; γ_s is the specific weight of sediments (kg m^{-3})

g is acceleration by gravity (m s^{-2})

A" is the experimentally obtained constant; A" = 0.047

B" is the experimentally obtained constant; B" = 0.25

G_s is the weight of sediments transported per unit of bed width per second (kg m^{-1} s^{-1})

The relation (4.25) applies for slope I=0.4 - 20‰, grain size d = 0.0004 to 0.03 m and depth of water h = 0.01 to 1.2 m at specific flow rate q_s = 0.002 to 2.0 m^3 m^{-1} s^{-1}.   Equation (4.25) may be written

$$y = A" + B"x = 0.047 + 0.25x \tag{4.26}$$

With a good knowledge of the basic hydraulic values of the flow profile and having determined the mean average diameter of sediment grain d_m and the specific gravity of the sediments it is possible to determine the slope gradient at which neither scouring nor settling occur. Nomogram in Fig. 58 will be helpful for the solution of equation (4.25).

The development of the stream bed does not end with the attainment of the compensated gradient. The compensation of the longitudinal gradient of the bed as well as the treatment of the bed and stream perimeter result in a decrease of the amount of transport sediments and in turn in the decrease of the sediment saturation of the water. This is followed by an increase in flow rate and tangential stress which in turn resumes the further movement and transport of sediments.

At a certain relation between gradient and the amount and depth of water, a state of equilibrium is reached at which water no longer erodes the stream bed. This state of equilibrium, which is often termed the equilibrium gradient, is minute and usually forms about 1/10 of the compensated gradient.

4.3 SURFACE RUNOFF MODELLING

Surface runoff modelling from catchment areas and from slopes has of late become a widely used practice. Mathematical models may be classified into:

- stochastic-conceptual

- stochastic-empirical

- deterministic-conceptual

- deterministic-empirical.

All these models may be either linear or non-linear. In stochastic models one variable is considered as being of random value to which pertains a certain distribution of probabilities. Deterministic models are characterized by the assumption that none of the variables is affected by random changes. Conceptual models are based on theoretical laws, empirical models on experiment or observation.

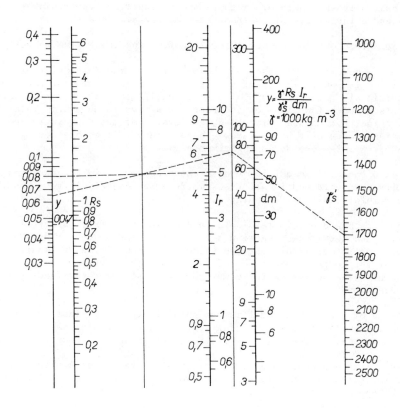

Fig. 58. Nomogram for the solution of equation (4.25).

The principle of superposition applies for the linear model which means that the sum of two input variables $x_1(t)$, $x_2(t)$ corresponds to that output which equals the sum of outputs $y_1(t)$, $y_2(t)$ pertaining to both inputs and that input $\alpha x_1(t)$ corresponds to output $\alpha y_1(t)$ where α is real number.

4.3.1 Models of Surface Runoff from Catchment Area

Mathematical models of direct runoff (flow rate and course of runoff wave) are usually based on unit hydrograph and on an instantaneous unit hydrograph. In this case function $w(t)$ expresses the instantaneous unit hydrograph in the given catchment area and function $x(t)$ the course of effective rainfall, the course of the direct runoff is expressed by function $y(t)$ which is given by the convolution integral.

$$y(t) = \int_{o}^{t} u(t-\tau)\, x\,(\tau)\, d\tau \qquad (4.27)$$

where t is time

τ - value from $<\sigma;t>$

From the theory of unit hydrograph and from the form of the convolution integral, which is basically the operator on the space of functions comprising the function of inputs and outputs, it follows that such models are linear.

A number of methods have been suggested to define the instantaneous unit hydrograph.

T. O'Donnell's model[34] is based on the measured course of precipitation and run-off from the catchment. The first step is the determination of the hyetograph of effective rainfall and hydrograph of direct runoff, then the first convolution coefficients of both functions of the Fourrier series in an appropriately chosen interval. From them may be determined the first p of the convolution coefficients of the instantaneous unit hydrograph. The accuracy of the obtained hydrographs of direct runoff mainly depends on the given number p.

Applying T. O'Donnell's model in several catchment areas in Czechoslovakia, relatively good results were obtained in all cases where an adequately large p was chosen and effective rainfall correctly separated.

Matrix model is based on discrete notations of functions $x(t)$ and $y(t)$, i.e. from vectors $\vec{x}, \vec{y}$ and determines function $u(t)$, represented by vector $\vec{u}$. By applying the convolution integral for discrete values a system of linear equations may be obtained for the sought coordinates of vector $\vec{u}$. This system has more equations than unknowns. The sought vector $\vec{u}$ may unambiguously be determined applying the least squares method.

The advantage of the matrix model is the possibility of eliminating inaccuracies in the separation of effective rainfall, i.e., in determining vector $\vec{x}$ using the integration method, it does, however, place high demands on computer memory.

J. E. Nash[27] in his model uses the analogy of a cascade of several reservoirs for determining function u (t). All reservoirs have the same outflow coefficient K to which applies

$$q = \frac{1}{K} V \tag{4.28}$$

where q is the outflow rate

     K is the outflow coefficient

     V is the volume of water in the reservoir.

Mathematical induction may be used to prove that instantaneous unit hydrograph of n reservoirs may be written

$$u(t) = \frac{1}{K} \frac{1}{\Gamma(n)} \left(\frac{t}{K}\right)^{n-1} \exp\left(-\frac{t}{K}\right) \tag{4.29}$$

where t is time

     n is the number of reservoirs

     K is the outflow coefficient.

$$\Gamma(n) = \int_o^\infty e^{-t} t^{n-1} dt$$

Equation 4.29 applies for any real positive n and K. J. E. Nash proved that this function u(t) may be considered as the equation expressing the instantaneous unit hydrograph. Equation (4.29) has only two constants (n and K) which are dependent on the given catchment. Nash evaluated them by the method of statistical moments. He also gives empirical formulas from which n and K may be determined on the basis of the main parameters of the catchment area (area, average gradient, length of main flow).

Very significant for erosion control is the modelling of the long-term runoff from
the catchment. Such models are known as balancing models. They are used to model
hydrological processes which take place in the catchment area as a complex of sub-
models. The sum of the effects of these sub-models make up the resulting output.
A. Becker's[3] model considers three components of runoff from the catchment area:

- surface runoff as the rapidly outflowing proportion of direct runoff

- surface runoff as the delayed proportion of direct runoff

- basic runoff as underground runoff from the catchment which flows out with
  a considerable delay, even in dry seasons.

The first component of runoff is related to the soil surface and to the hydrographic
system, the second to soil (layers of higher conductivity) and the third to the
ground water system.

Sub-models are of two types:

- sub-models generating runoff; they determine the proportion of precipitation
  retained in the catchment and the proportion of the evapotranspiration and
  thereby the proportion of precipitation forming proper runoff

- sub-models of runoff concentration; they simulate the time delay of runoff
  generated by previous sub-models.

The individual sub-models may be constructed using various linear and non-linear
reservoir cascades. Becker used a computer for computing the outputs of the individual sub-models, the logical operation dependent on these outputs and the calculation of total runoff.

Czechoslovak authors of runoff models include J. Balek and L. Jokl[2] who expressed
the relation between effective rainfall x(t) and the hydrograph of direct runoff
y(t) by a linear differential equation of the second order with constant coefficients
A,B,C written

$$A \frac{d^2y}{dt^2} + B \frac{dy}{dt} + Cy - x(t) \qquad\qquad (4.30)$$

on condition that

$$\int_o^T y(t)dt = K \int_o^T x(t)dt \qquad\qquad (4.31)$$

where $K \in (0,1)$.

The model was demonstrated on the Volyňka river catchment. A,B,C,K are parameters
of the model. The authors used the Laplace transformation to discuss these parameters; in the expression of function y they arrived at a convolution integral with
a function expressing the instantaneous unit hydrograph in dependence on parameters
$A,\bar{B},\bar{C}$ ($K = \frac{1}{C}$). For determining the respective parameters the authors used the
moment method and by taking an average they obtained the desired values for the
observed catchment.

4.3.2 Model of Surface Runoff from Slope

This model has been constructed by M. Holý and J. Mls[14] using the following values:
the parameters of the slope (gradient, length), soil properties (initial soil
moisture, infiltration), and the intensity and time course of precipitation.

For the mathematical expression of the basic relations they assumed that:

- the surface of the slope is a plane forming angle α with the horizontal plane
- the slope width is unlimited
- the intensity of rainfall water is even on the whole slope and is only a function of time
- the infiltration rate of water in the soil is only a function of time (Fig. 59).

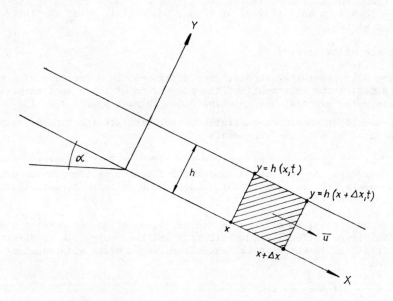

Fig. 59.   Conditions for constructing model of surface runoff·
          after M. Holý and J. Mls.

The laws on the conservation of mass and momentum and the equation of continuity and of motion apply for water running off from the surface of the slope. The authors derived the following equations for surface runoff from slopes:

the equation of continuity

$$\frac{\partial(\overline{u}h)}{\partial x}(x,t) + \frac{\partial h}{\partial t}(x,t) = r'(t) - i'(t) \tag{4.32}$$

the equation of motion

$$h(x,t) \frac{\partial \overline{u}}{\partial t} (x,t) + h (x,t) \overline{u} (x,t) \frac{\partial \overline{u}}{\partial x} (x,t) = g \sin \alpha h (x,t) -$$

$$-g \cos \alpha h (x,t) \frac{\partial h}{\partial x} (x,t) - g \cos \alpha h*(t) \frac{\partial h}{\partial x} (x,t) - \tag{4.33}$$

$$-\frac{\tau(h,\overline{u})}{\rho} + r'(t)v*(t) \sin\alpha$$

where h is the height of surface runoff as a function of place and time (m)

α is the slope gradient ($^{\circ}$)

$\overline{u}$ is the mean flow velocity of surface runoff in the direction of axis X
    (m s^{-1})

r' is intensity of precipitation (m s^{-1})

i' is intensity of infiltration of runoff water into the soil (m s^{-1})

t is time (s)

g is acceleration by gravity (m s^{-2})

h* is the height determining the increment of pressure caused by raindrops
    falling on the soil surface (m)

τ is tangential tension (N m^{-1})

ρ is water density (kg m^{-3})

v* is mean velocity of raindrop fall (m s^{-1}).

From the equations of continuity and of motion M. Holý and J. Mls derived a differential equation for a stationary case which may be written

$$\frac{d\overline{u}}{dx} = \frac{g \sin\alpha \ (r'-i')x\overline{u}^2 - g \ \cos\alpha(r'-i')^2 x\overline{u} - g\cos\alpha \ h*(r'-i')\overline{u}^2}{(r' - i')x\overline{u}^3 - g \ \cos\alpha(r'-i')^2 x^2 - g \ \cos\alpha h* \ (r'-i')x\overline{u}} \tag{4.34}$$

$$- \frac{C(r'-i') \ \partial_x \partial_{\overline{u}} 3 + \beta - \partial + \ r'v* \ \sin\alpha \overline{u} \ 3}{(r'-i')x\overline{u}^3 - g \ \cos\alpha(r'-i')^2 x^2 - g \ \cos\alpha \ h*(r'-i')x\overline{u}}$$

The solution of the above equation will determine the depth and velocity of surface runoff and thereby also the value of surface runoff. A finite differential method may be used to solve the equation.

4.4   TRANSPORTATION PROCESSES

During erosion processes substances are transported due to the energy of erosion factors, especially water and wind. Data on the transportation of substances by the force of wind are numerous and have been derived from aerodynamics and tried and proved in experimental aerodynamic tunnels. The transportation of substances over the ground surface and in the soil by water pose a more difficult problem which still remains to be solved. These transportation processes are extremely complex mainly owing to the fact that during transportation the substances mix and are affected by the medium which results in various physical, chemical and biological processes taking place and making the transportation result very difficult to enumerate. The study of problems related to transport and the application of current knowledge to the transportation of substances by water are therefore very important for ascertaining the consequences of erosion.

4.4.1 *Fundamental Principles of the Mechanism of Transportation Processes*

In the study of transportation processes the most important factors are the transmission of momentum (the flow of viscose fluids), the transmission of energy (heat conduction, convection, radiation) and the transmission of matter (diffusion).

4.4.1.1 Transmission of Momentum

The resolution of the transmission of momentum in flowing viscose fluids is Newton's law of viscosity in the form

$$\tau_{yx} = - \mu \frac{dv_x}{dy} \qquad\qquad (4.35)$$

where τ_{yx} is shear stress (Pa)

v_x is the component of the vector of velocity along the X-coordinate (m s^{-1})

μ is the dynamic viscosity of the fluid (Pa s).

The fluids which behave according to this law are termed Newtonian fluids.

It follows from relation (4.35) that the viscosity of the flow of momentum takes the direction of the negative gradient of velocity from which it follows that momentum takes the direction of decreasing velocity.

The most important application of Newton's viscosity law for the transportation processes of water erosion is the resolution of the distribution of velocity in laminar flow in systems of simple geometrical configurations. From this it is possible to obtain maximum velocity, average velocity and tangential stress at the surface. The solution is based on the balance of momentum in a thin layer of the fluid. It applies for the state of equilibrium that

> the inflow of momentum – the outflow of momentum + the total of the
> forces affecting the system = 0   (4.36)

The laminar flow of a purely isothermal fluid may be expressed by the basic equations on the conservation of mass and momentum. Further equations are required for solving the flow of non-isothermal fluids and multi-component fluid mixtures. Such equations should express the conservation of energy and the conservation of mass of the individual chemical components.

4.4.1.2 Transmission of Energy

The transmission of energy may take place in the presence of mass conduction when energy transmission occurs by conduction and convection or in space which is void of substances. In the latter case energy transmission takes place by radiation.

The transmission of thermal energy is conditional upon the inequality of temperature in neighbouring points of the material environment; in such a case heat is transmitted by conduction. In the existence of a liquid which has the capability of moving freely and carrying energy with it heat transmission by convection takes place. The transmission of heat by radiation takes place by electromagnetic wave motion whose speed equals the velocity of light.

The basic equation for heat transmission by conduction is the Fourier principle which expressed in a one-dimensional form may be written

$$q_y = - k \frac{dT}{dy} \qquad\qquad (4.37)$$

where q_y is the density of the heat flow in the positive direction (W m^{-2})

   k is heat conductivity (W m^{-1} K^{-1})

   T is absolute temperature (K).

The density of the heat flow is thus proportional to the temperature gradient; k has similar significance as μ in the transmission of momentum (4.35).

For an equilibrium state of the system the law on the conservation of energy applies for the selected layer whose surfaces are perpendicular to the direction of the heat conduction and may be written:

> velocity of supply _ velocity of withdrawal
> of thermal energy   of thermal energy   +
>
> +   velocity of production
>   of thermal energy   = 0   (4.38)

Where the thickness of the layer approximates zero a differential equation is obtained for heat distribution.

The solution of the transmission of energy is considerably complicated in dependence on the introduction of initial and boundary conditions.

4.4.1.3   Transmission of Mass

The fundamental law for the transmission of mass (diffusion) is Fick's law in the form

$$Ja = - cD_{AB}\nabla x_A \qquad (4.39)$$

where J_a is the density of the diffusion flow of the mass (m^{-2} s^{-1})

   c is total molar concentration (mol m^{-3})

   D_{AB} is the binary coefficient of diffusion for the system from components A
        and B (m^2 s^{-1})

   x_A is the molar fraction of component A (dimensionless).

Relation (4.39) expresses the fact that the density of the diffusion flow of component A is proportional to the concentration gradient.

In the one-dimensional system the relation may be written

$$J_{ay} = - D_{AB} \frac{d}{dy} (\rho_A) \qquad (4.40)$$

where ρ_A is the concentration of component A (kg m^{-3}).

For the investigation of diffusion in liquids it is possible to apply the hydro-dynamic theory following from the Nernst-Einstein equation written

$$\overline{D}_{AB} = \overline{k}T \frac{u_A}{F_A} \qquad (4.41)$$

where $\overline{D}_{AB}$ is the diffusivity of one particle of the molecule of dissolved mass A
        in inert medium B (m^2 s^{-1})

   $\frac{u_A}{F_A}$ is the balancing speed which a particle of mass A has attained by the
        effects of unit force (m s^{-1} N^{-1})

   $\frac{T}{k}$ is absolute heat (K)
        is Boltzmann's constant (J K^{-1}).

From the hydrodynamical model it is possible to obtain the diffusion coefficient for spherical molecules in a diluted solution and the coefficient of self-diffusion. The hydrodynamic theory shows that the form of the diffusing particle affects the value of the diffusion coefficient owing to a change taking place in the friction coefficient.

Diffusion, like the transmission of momentum and energy, may be resolved applying the law on the conservation of mass which in this case may be expressed

velocity of supply _ velocity of withdrawal
of component A   of component A   +

velocity of production of
+   component A by homogenous   = 0   (4.42)
chemical reaction

Component A may enter or depart from the system by diffusion as well as by the flow of the fluid as a whole.

The construction of a linear equation for an infinitely small thickness of layer will yield a differential equation whose solution is the distribution of component A in the system.

4.4.2 *Transportation Processes in Surface Waters*

The transportation processes of substances detached by erosion in the catchment have great significance for changing the quality of surface water resources. Of late considerable quantities of mineral fertilizers, pesticides etc. are being applied to the soil which pollute surface and ground water supplies and are the cause of a significant eutrophication of water in reservoirs. Attempts have been made to construct models representing the transportation of these substances and allowing forecasts to be made of the resulting effects of transportation processes.

Dynamic models of transportation processes in surface waters worthy of attention include the dynamic model constructed by K. J. Dahl-Madsen and E. Gargas[9]. The model derived from the eutrophication of shallow fjords may be applied to water reservoirs in which complete vertical mixing of water takes place.

Transportation and changes in the concentration of substances in the horizontal direction have been expressed by introducing the system of turbulent water units (Fig. 60).

Figure 60 shows that the transportation of substances between each turbulent water unit of the system takes place by two different mechanisms, namely advective transportation is induced by the mere transportation effect of water in the direction of the flow, mixing transportation is induced by the continuous mixing of water particles of different concentrations. Advective transportation in the system is a function of the inflow of water from the catchment, mixing transportation is a function of the water turbulence.

For an arbitrary substance in an i-th turbulent unit of water the law on the conservation of mass may be expressed in the form

$$V_i \frac{dC_i}{dt} = q_{i-1,i}C_{i-1} + q_{i,i+1} C_{i+1} - (q_{i-1,i} + q_{i,i+1}) C_i + \quad (4.43)$$

$$+ Q_{i-1,i}C_{i-1} - Q_{i,i+1}C_i$$

where V is the volume of the turbulent unit of water (m³)

Q is the inflow of pure water (m³day⁻¹)

q is the transportation of water mixed with substances (m³ day⁻¹)

C is the specific concentration (g m⁻³).

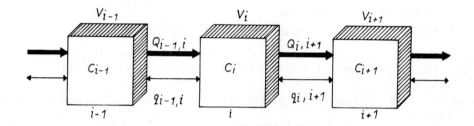

Fig. 60.   System of turbulent water units for ascertaining
the distribution of the concentration of substances
in water.

From equation (4.43) time changes in the specific concentration of the substance
in turbulent water units in dependence on changes in the volume can be calculated.
Concentration varies not only with the amount of substances flowing in from the
catchment but also with their participation in the biological and chemical reactions
taking place in the water and by their sedimentation. As all necessary input data
required for the general application of the model are not yet available the authors
recommend that it should only be used in such cases where at least some measurements
enabling confrontation have been made.

With regard to the complexity of dynamic models for which it is difficult to determine input and boundary conditions statistical models are often used.

4.4.3 *Transportation Processes in Soil*

Water enriched by erosion processes with various substances infiltrates into the
soil and often reaches the level of the groundwater table. The flow in which
greater quantities of substances are gradually mixed with water is described as
miscible flow.

4.4.3.1 Dispersion of Substances in Soil

During the course of the infiltration into the soil of substances dissolved in water
a change takes place in the soil profile, both in the sideward and downward directions of the concentration of substances in the water owing to their dispersion in
the soil. The dispersion is affected by the porosity of the soil, the aggregation
of particles, the properties of flowing water, the conditions of the flow, etc.
Dispersion differs depending on whether it is longitudinal or transverse to the
flow and it is characterized by the coefficient of transverse and longitudinal
dispersion.

Dispersion is effected by hydrodynamical factors and molecular diffusion[22].
Hydrodynamical dispersion may be distinguished from molecular diffusion applying
the Péclet number written

$$Pe = \frac{v}{P_p} \frac{d_a}{D_m} \qquad (4.44)$$

where v is the macroscopic velocity of the water flow in the soil (m s^{-1})

P_p is soil porosity (dimensionless)

d_a is mean grain diameter (m)

D_m is the coefficient of molecular diffusion (m^2 s^{-1}).

It approximately applies that at Pe < 3.10^{-1} molecular diffusion is effective,
when Pe = 3.10^{-1} to 5 molecular diffusion and hydrodynamical dispersion have the
same significance

when Pe = 5 to 20 an interference occurs between molecular diffusion and hydrodynamical dispersion
when Pe > 20 molecular diffusion is not taken into consideration.

The empirical relation between the value of dispersion coefficient D and the velocity of the flow[32] shown in Fig. 61 shows that molecular diffusion should not
be ignored in resolving problems related to miscible flow in the soil.

Longitudinal dispersion which is important for the ascertainment of the extent of
soil profile pollution may be ascertained from the penetration curve which may be
determined experimentally[21]. Penetration curves of the water flow through the
soil comply with Gause's normal distribution.

4.4.3.2 Basic Equation for the Transportation Process in Soil

It follows from the law on the conservation of mass that a change in the concentration of substance at a given point equals the difference between its inflow and
outflow. For the one-dimensional process it applies that

$$\frac{\partial C}{\partial t} = - \frac{\partial q}{\partial x} \qquad (4.45)$$

where C is concentration (kg m^{-3})

q is the flow rate transversely to the cross-sectional area within a unit
of time (kg m^{-2} s^{-1})

t is time (s)

x is the ordinate (m).

Equation (4.45) applies for water saturated media. In unsaturated media the content of the substance changes not only with changes of concentration but also with
changes of soil moisture and the equation (4.45) may be written in the form

$$\frac{\partial(\theta C)}{\partial t} = - \frac{\partial q}{\partial x} \qquad (4.46)$$

where θ is moisture (m^3 m^{-3})

Changes in flow rate result from dispersion and convection. It applies for dispersion flow q that

$$q = -D \frac{\partial(\theta C)}{\partial x} \tag{4.47}$$

where D is the diffusion coefficient (m^2 s^{-1}) for convection (the transportation of substances by liquid flow)

$$q = vC \tag{4.48}$$

where v is the macroscopic (Darcy's) rate of flow of water through the soil.

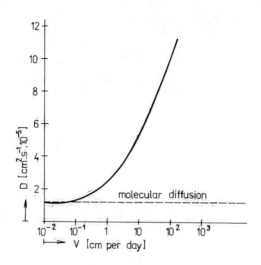

Fig. 61. Relation between dispersion coefficient D and
flow velocity after D. R. Nielson.

The combination of equations (4.46) (4.47) and (4.48) will be applied to obtain the basic equation of dispersion in the unsaturated soil medium written

$$\frac{\partial(\theta C)}{\partial t} = \frac{\partial}{\partial x}[D \frac{(\theta C)}{\partial x}] - \frac{\partial(vC)}{\partial x} \tag{4.49}$$

For the saturated soil medium the dispersion equation may be written in the form

$$\frac{\partial C}{\partial t} = D* \frac{\partial^2 C}{\partial x^2} - v_p \frac{\partial C}{\partial x} \tag{4.50}$$

where v_p is mean pore velocity (m s^{-1})

D* is the diffusion coefficient (m^2 s^{-1}).

These equations apply on condition that the soil is inert to substances contained in the water and that the physical properties of the water containing the substances do not change with the change of concentration.

Where processes are taking place in the soil in which substances participate which
are contained in the infiltrating water (dissolution, precipitation, ion exchange,
microbiological processes, etc.) and where substances cause changes in concentration,
equation (4.50) should be extended to the following equation which may be written

$$\frac{\partial C}{\partial t} = D* \frac{\partial^2 C}{\partial x^2} - v_p \frac{\partial C}{\partial x} + \Sigma \, f \, n(C,x,t) \tag{4.51}$$

where n is the type of process taking place in the soil.

The solution of this equation is extremely complicated requiring knowledge and the
expression of all processes taking place in the soil, i.e., biological, chemical
and other complex changes in substances transported by water, such as occur on
the substances, contact with the soil.

4.4.3.3 Transportation of Nitrogen in Soil

The transportation of nitrogen in the soil which was studied by H. E. Jensen[18]
serves as a good example of the application of the theory discussed in the previous
chapter.

The transportation of nitrogen in the soil results from the convective flow of
infiltrating water into the soil and the molecular diffusion of nitrogen dissolved
in the water owing to the gradient of concentration. The two processes may take
place concurrently in the same or opposite direction.

Next to diffusion there is hydrodynamical dispersion caused by differences in the
flow velocity through soil pores and changes in mean flow velocity owing to the
different soil pores. At sufficiently high velocities hydrodynamical dispersion
may considerably exceed the effects of diffusion. In such a case the diffusion
coefficient D* in equation (4.50) should be viewed more as the coefficient of
dispersion than as the coefficient of diffusion.

On these assumptions we shall obtain for the calculation of concentration the equation after W. G. Gardner (11) written

$$C = \frac{C_o z_o}{\sqrt{4_\pi D* t}} \; \exp - \frac{(z - vt)^2}{4D* t} \tag{4.52}$$

where C_o is concentration at a small depth z_o.

Equation (4.52) assumes that the average flow velocity of water through the pores
and that of nitrate ions is identical which need not necessarily apply at low ion
concentrations in soils with a high ion exchange capability.

Maximal concentration C_{max} as a function of depth is expressed in the equation

$$C_{max} = \frac{C_o z_o}{\sqrt{2\pi \beta z_{max}}} \tag{4.53}$$

where β is the index of dispersion; $\beta = \frac{2D*}{v}$.

Equation (4.53) may be applied to obtain the index of dispersion and the coefficient
of dispersion when values C_o, z_o, C_{max}, z_m, v are known.

In accordance with equation (4.53) maximum concentration drops at a constant velocity. The narrow strip of nitrates expands to cover a greater part of the soil

profile owing to diffusion and hydrodynamic dispersion and the nitrate concentration tends to reach Gause's distribution with a standard deviation $\sqrt{2D*t}$. Dispersion proceeds and the concentration gradient decreases with time t.

H. E. Jensen[18] proved these equations by field experiments and arrived at the conclusion that they may be applied for the investigation of the transportation of substances in soil. This is shown in Fig. 62 which gives the concentration $NO_3 - N$ in soil water in dependence on depth according to equation (4.52).

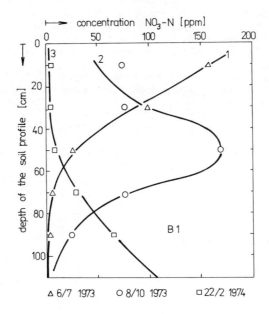

Fig. 62. $NO_3 - N$ concentration in soil water (ppm) after H. E. Jensen.

Curve 1 represents the relation on one day of the vegetation season with no rainfall, curve 2 shows the relation at an infiltration of 80 mm of water and curve 3 at an infiltration of 400 mm of water. The shape of the obtained curve corresponds to equation (4.52).

REFERENCES

1. Balek, J. and Jokl, L., *Rainfall-Runoff Linear Deterministic Model of the Second Order*, Institute of Hydrology, Czechoslovak Academy of Sciences, 325/9/73, 1973.
2. Balek, J. and Jokl, L., *Rainfall-Runoff Deterministic Linear Model of Second Order*, Journal of Hydrological Sciences 2, No.3-4, 1974.
3. Becker, A., The Integrated Hydrological Catchment Model EGMO, *Symposium or Application of Mathematical Models in Hydrology*, Bratislava, 1975.
4. Benetin, J., *Dynamics of Soil Moisture*, Slovak Academy of Sciences, Bra 1970.

5.   Bird, R. B., Stewart, W. E. and Lightfoot, E. N., *Transmission Processes*,
     Academia, Praha, 1968.
6.   Bresler, E., Simultaneous Transport of Solutes and Water under Transient
     Unsaturated Flow Conditions, *Water Resources*, Vol.4, 1973.
7.   Cablík, J. and Jůva, K., *Soil Conservation*, Praha, 1963.
8.   Clarke, R. T., Mathematical Models in Hydrology, *Irrigation and Drainage Paper*
     No.19, FAO, Rome, 1973.
9.   Dahl-Madsen, K. I. and Gargas, E., A Preliminary Eutrophication Model of
     Shallow Fjords, *Proceedings 7th International Conference on Water Pollution*,
     Paris, 1974.
10. Gardner, W. R., Solutions of the Flow Equation for the Drying of Soils and
     other Porous Media, *Soil Sci.Soc.Amer.Proc.* 23, 1959.
11. Gardner, W. R., Movement of Nitrogen in Soil, Soil Nitrogen, *Agron.* No.10, 1965.
12. Garslaw, H. S. and Jaeger, J. C., *Conduction of Heat in Solids*, London, 1959.
13. Gončarov, V. N., Osnovy dinamiky ruslovykh procesov, *Gidrometeoizdat*, Leningrad,
     1954.
14. Holý, M. *et al.*, *Building up of the Mathematical Model of Surface Runoff with
     Regard to its Erosion Effects*, Report on Research Program, Prague, 1975.
15. Chen Cheng Lung and Chow Ven Te, *Hydrodynamics of Mathematically Simulated
     Surface Runoff*, University of Illinois, 1968.
16. Chow Ven Te, *Handbook of Applied Hydrology*, McGraw-Hill Book Co., New York,
     London, 1964.
17. Chow Ven Te, Laboratory Study of Watershed Hydrology, *Proc.Int. Hydrology
     Symposium*, University Ft Collins, 1967.
18. Jensen, H. E., Nitrogen Movement and Leaching in Soil, *Nordic Hydrology*, 1976.
19. Karantounias, G., Dünnschichtabfluss auf stark geneigter Ebene, *Mitteilungen*,
     Heft 192, Universität Karlsruhe, 1974.
20. Kirkham, D. and Powers, W. L., *Advanced Soil Physics*, New York, 1972.
21. Kutílek, M., Steady Vertical Flows above Groundwater Table, *Proceedings IX.*
     Congress ICID, New Delhi, 1975.
22. Kutílek, M., *Soil Science*, SNTL, Prague, 1978.
23. Levi, I. I., *Dinamika russlovykh potokov*, Moskva-Leningrad, 1948.
24. Mayer-Peter, E. and Müller, R., Formulas for Bed-Load Transport, Stockholm,
     Proceedings Third Meeting IAHR, 1948.
25. Mezentsev, V. S., K. teorii formirovaniya poverkhnostnogo stoka so sklonov,
     Meteorologiya i gidrologiya 3, 1969.
26. Mircchulava, C. E., *Inzhenyerniye metody rascheta i prognoza vodnoy erozii*,
     Moscow, 1970.
27. Nash, J. E., The Form of the Instantaneous Unit Hydrograph, *Intern.Assoc.Sci.
     Hydrology*, Pub. 45, 1957.
28. Nash, J. E., A Unit Hydrograph Study with particular Reference to British
     Catchments, *Proc. Inst. Civil Eng.*, 1960.
29. Němec, J., *Hydrology*, SZN, Prague, 1965.
30. Nyerpin, S. V. and Chudnovskiy, A. F., *Fizika pochvy*, Moscow, 1967.
31. Nielsen, D. R. and Biggar, J. W., Miscible Displacement, III Theoretical
     Considerations, *Soil Sci.Proc.*, Vol.25, 1962.
32. Nielsen, D. R., Jackson, R. D., Cary, J. W. and Evans, D. D., Soil Water,
     Am.Soc.Agron., Madison, 1972.
33. Nielsen, K. S. and Nyholm, N., *The Contribution of Nutrient from Diffuse
     Sources*, Water Quality Institute, Denmark, 1976.
34. O'Donell, T., Instantaneous Unit Hydrograph Derivation by Harmonic Analysis,
     Symp. on Underground Water, IASA Pub., Athens, 1960.
35. *Proceedings of the Symposium on Mathematical Models in Hydrology*, SAV, Bratis-
     lava, 1975.
36. Schröder, H., Fjordmodelling, 9th Inter-nordic Water Research Conference,
     Trondheim, 1973.
37. Valentini, C. D., *Del modo di determinare il profilo compensazione e sua impor-
     tanza nelle sistemazioni idrauliche*, Milano, 1895.

38. Velikanov, M. A., *Gidrologiya sushi*, Leningrad, 1964.
39. Waklhu, O. N., An Experimental Study of Thin-Sheet Flow over Inclined Surfaces,
     Mitteilungen, Heft 158, Universität Karlsruhe, 1970.

5. Predicting Intensity of Water Erosion and Modelling Erosion Processes

Erosion is caused by rainfall and by surface runoff and is affected by a number of natural and anthropogenic agents. It may be expressed as the relation between the erosivity of rainfall, i.e., the potential ability of rain to cause erosion, and soil erodibility, i.e., the susceptibility of the soil to erosion.

A number of authors have tried to ascertain this relation from field observations, field experiments, laboratory experiments and theoretical studies. Rain as the principal erosion agent was usually characterized by intensity, in some cases by intensity, the raindrop size and raindrop velocity, soil properties were expressed by coefficients showing the effects of soil texture and structure on the soils' infiltration capability and by other factors affecting the origination and course of erosion processes, namely slope gradient, slope length, the vegetative cover, etc. Scientists have attempted to interrelate these factors with the aim of obtaining erosion intensity expressed by soil loss from a soil unit over a unit of time.

5.1 EROSION INTENSITY IN DEPENDENCE ON SURFACE RUNOFF VELOCITY AND ITS TANGENTIAL STRESS

The most frequently used expressions are those which express erosion intensity in dependence on surface runoff velocity and its tangential stress. This is because these values are most frequently monitored by hydrologists. An example of this procedure is the equation derived by G. I. Shvebs[22] and written

$$S_p = adv \left(\frac{\bar{v} - v_o''}{v_o''} \right)^3 \tag{5.1}$$

where S_p is the rate of soil loss (kg s^{-1})
  d is the mean diameter of detached particles (mm)
  v is the velocity of runoff (m s^{-1})
  $\bar{v}$ the energy parameter dependent on the energy of raindrops and the velocity of surface runoff
  v_o'' is the initial value of the energy parameter at which the detachment of particles of the given size is initiated
  a is the coefficient of proportionality.

The important factor in all equations written by various scientists is the knowledge of maximum permissible runoff velocity and maximum tangential stress.

After C. E. Mirckhulava[17] maximum permissible runoff velocity and maximum
tangential stress result in what he terms permissible soil loss tolerance, at
which the loss of soil particles is replaced by the annual formation of new soil
particles. As an example he recalls data obtained by H. Kohnke and A. Bertran[13]
who have calculated that under certain conditions 6.75 t ha^{-1} of new soil are
formed annually.

After C. E. Mirckhulava the erosion limit at which erosion control measures must
be taken occurs at surface runoff velocity

$$v_x > 1.15 \ v_p \qquad\qquad (5.2)$$

where v_x is runoff velocity at distance x from the water divide
      v_p is the maximum permissible runoff velocity.

M. A. Velikanov[24] writes the following relation for permissible runoff velocity

$$v_p = 3.14 \ \sqrt{15d + 0.006} \qquad\qquad (5.3)$$

where v_p is the extreme permissible surface runoff velocity (m s^{-1})
      d is the mean grain diameter (m).

The permissible velocity of surface runoff from different soils may be assessed
using the mean grain size characteristic D_{50} (Table 5.1).

TABLE 5.1.   Permissible velocity of surface runoff in
             dependence on D_{50}

Content of particles of 1st cat.(%) grain size smaller than 0.01 mm	Soil	D_{50} (mm)	v_p (m.s^{-1})
0-10	sandy soil	0.55	0.37
10-20	loamy sand	0.20	0.31
20-30	sandy loam	0.12	0.28
30-45	loamy soil	0.03	0.26
45-60	clay-loamy soil	0.01	0.25
60-75	clay soil	0.002	0.24
>75	loam	0.001	0.24

I. I. Levi[15] obtained for concentrated runoff an equation for the quadrat area
of flow at d>1.5 mm written

$$v_p = 1.3\sqrt{gd} \ (0.8 + \frac{2}{3} \ \lg \ \frac{10R}{d_k}) \ [cm.s^{-1}] \qquad\qquad (5.4)$$

for the transition area of flow

$$v_p = 35d^{0.25} \ (\lg \ 7.5 \ \frac{R}{d} - 6d) \ [\,cm.s^{-1}] \qquad\qquad (5.5)$$

for smooth beds (d<0.25 mm)

$$v_p = \frac{100R^{1/8}}{\sqrt{7.5 + R^{\frac{1}{4}}}} \qquad\qquad [\,cm.s^{-1}] \qquad\qquad (5.6)$$

where d is the mean grain diameter (cm)
      g is acceleration by gravity (cm s^{-2})
      d_k is the diameter of the biggest fractions which make up 10% of the
         sample (cm)
      R is the hydraulic radius (cm).

The informative values of permissible velocities and permissible tangential stress
for concentrated runoff are given in Tables 4.2 and 4.3. Informative values of
permissible runoff velocity and tangential stress for sheet surface runoff have
been obtained by E. Dýrová (Fig. 63)[5].

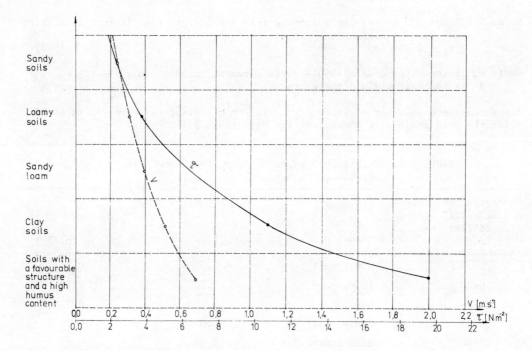

Fig. 63. Informative values of permissible v and τ for
sheet surface runoff after E. Dýrová.

5.2 EMPIRICAL MODELS OF EROSION PROCESSES

The erosion process is most frequently expressed by the relation between resulting
intensity determined by the gravity or volume of soil loss from a unit area over a
given time unit and the erosion agents. The general relation may be written as

$$S_P = f(X_K, X_H, X_M, X_S, X_G, X_V, X_T, X_{EK})\tag{5.7}$$

where S_P is erosion intensity (soil loss)
     X_K is the climatic factor

X_H is the hydrological factor
X_M is the morphological factor
X_S is the soil factor
X_G is the geological factor
X_V is the vegetative factor
X_T is the technical factor
X_{EK} is the socio-economic factor

A detailed analysis of these factors will show that their interactions are extremely complex and that it is difficult to model the erosion process. This is mainly because the interactions and effects of the individual factors are manifest in extremely varied conditions.

On the basis of the study of erosion factors and their interactions M. Holý[11] arrived at the conclusion that the equation (5.7) may be transformed to the equation written

$$S_p = f(a, X_H, S', L')$$ (5.8)

where X_H is the hydrological factor
S' is the factor of the slope gradient
L' is the factor of the slope length
a is the coefficient of the effect of other factors which play a part in the erosion process, insofar as they are not included in X_H.

The relationship between the factors was investigated and proved on field experimental plots with a 44.5% gradient and a slope length of 19.8 m. On the basis of a 15-year observation period M. Holý obtained the following equation for a bare soil surface

$$S_p = aq^b$$ (5.9)

where S_p is soil loss (kg ha^{-1})
q is surface runoff from precipitation (m^3 ha^{-1})
a,b are coefficients dependent on local conditions; for the area of northern Bohemia the coefficients for sheet erosion are
a = 2.002,   b = 1.103, for rill erosion
a = 0.210,   b = 2.645.

The model of the erosion process for sheet erosion was constructed on the basis of the erosion intensity investigated on the slope profiles in the experimental area in northern Bohemia. M. Holý proceeded from the assumption that the amount of soil particles detached and washed out from the topsoil and their size, depend on the intensity of the erosion process, provided all other conditions are identical. It thus follows that changes in topsoil texture on slopes indicate the intensity of selective sheet erosion. On the basis of these findings he classified soil texture into 8 categories (Table 5.2).

For the soil texture given by this classification he determined the common characteristic expressed as

$$\mu = \alpha_1 m_1 + \alpha_2 m_2 + \ldots + \alpha_k m_k$$ (5.10)

where μ is the characteristic of the soil texture
$m_1, m_2 \ldots \ldots m_k$ is the weight percentage of the individual categories of soil grains
$\alpha_1, \alpha_2, \ldots \alpha_k$ are the inverse values of the respective sizes of soil particles R

$$\alpha_k = \frac{1}{R_k}$$

It thus applies that

$$\mu = \frac{1}{R_1} m_1 + \frac{1}{R_2} m_2 + \ldots + \frac{1}{R_k} m_k \tag{5.11a}$$

After substituting values R_1 M. Holý obtained the equation

$$\mu = \frac{1}{0.005} m_1 + \frac{1}{0.03} m_2 + \frac{1}{0.075} m_3 + \frac{1}{1.0} m_4 +$$

$$+ \frac{1}{3.5} m_5 + \frac{1}{6.0} m_6 + \frac{1}{11.0} m_7 + \frac{1}{22.5} m_8 \tag{5.11b}$$

TABLE 5.2. Characteristic grain size of soil particles

Category	Grain size (mm)	Characteristic grain size R (mm)
I	<0.01	0.005
II	0.01-0.05	0.03
III	0.05-0.10	0.075
IV	0.10-2.00	1.00
V	2.0-5.0	3.5
VI	5.0-7.0	6.0
VII	7.0-15.0	11.0
VIII	15.0-30.0	22.5

A number of investigations showed that the obtained value μ adequately characterizes the textural condition of the topsoil at different points of the slope.

The relation between characteristic μ of the site, which according to the investigations has suffered maximum soil loss (the place on the slope is determined by the lowest value μ) and characteristic μ at the slope water divide where a relatively constant textural condition is assumed, is described as the maximum relative change of topsoil texture. Detailed investigations of the course of erosion on tilled farmland slopes as related to maximum relative changes of texture proved that these changes are directly proportional to the intensity of the erosion process. The results of investigations carried out on 10 selected slopes are given in Fig. 64.

The model only applies for slopes with uniform tillage over a longer period of time and for slopes where field boundaries have been preserved.

When evaluating the effects of erosion factors on the resulting intensity of the erosion process some scientists have arrived at the conclusion that the erosion intensity is a function of erosivity of rainfall E_d and erodibility of soil E_p written as

$$S_p = f(E_d, E_p) \tag{5.12}$$

The graphical expression of this equation after N. Hudson[12] is shown in Fig. 65.

Great progress was made in this respect by the model of the erosion process con-structed by G. W. Musgrave[18] on the basis of the results of long years of erosion research and monitoring in the USA. His equation may be written as

$$S_p = K'C'I^{1.35} L'^{0.35} i_{30}^{1.75}$$ (5.13)

where S_p is soil loss over a one year period
      K' is the factor of the soil erodibility
      C' factor of the vegetative cover
      I is slope gradient
      L' factor of the slope length
      i_{30} is a 30 min rainfall with p = 0.5

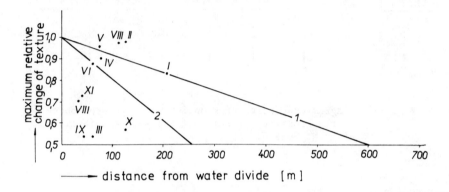

Fig. 64. Erosion intensity expressed by maximum textural
          changes on investigated slopes and their
          classification into three categories by the
          extent to which the soil is affected by erosion.

5.2.1 *W. H. Wischmeier and D. D. Smith's Empirical Model of Erosion*

W. H. Wischmeier and D. D. Smith[26,27] specified and extended equation (5.13)
on the basis of a wide range of observations of the erosion processes and the
effects of other factors. Their equation - the "universal soil-loss equation" -
is written as

$$S_p = R'K'L'S'C'P'$$ (5.14)

The effect of the individual members of the equation on erosion intensity is
assessed with regard to their effects on a unit plot with specified parameters:
the length of the unit plot is 22.13 m, slope gradient 9%, the plot is continually
fallow, and tilled up and down the slope gradient.

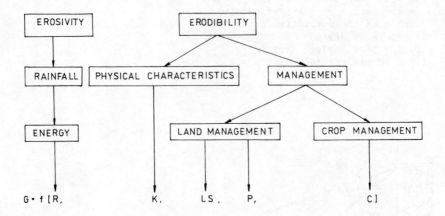

Fig. 65.   Diagram expressing erosion intensity in dependence
           on the erosivity of rainfall and soil erodibility
           after N. Hudson.

The individual members of the equation (5.14) are:

S_p is annual soil loss in t ha^{-1}; where the equation is applied for erosion
   control measures, S_p is the so-called soil loss tolerance, i.e., such soil
   loss which may be permitted in the implementation of erosion control measures
   without reducing the productivity of the plot

R' is the rainfall factor defined as the coefficient of the kinetic energy
   of rain and its maximum 30 min intensity; $R' = EI_{30}$

K' is the factor of the soil erodibility – expresses the susceptibility of
   the soil to erosion and gives the loss of soil particles from the unit plot
   per unit of rainfall factor R'. For the given soil it may be experimentally
   investigated from the relation $K' = S_p R'^{-1}$. The simple determination of
   factor K' may in case of a lack of experimental data be obtained from its
   dependence on five fundamental characteristics of the soil, i.e., the
   content of soil particles (<0.10 mm), the content of soil particles
   (0.10–2.00 mm), the content of organic substances (% of humus), soil structure
   and soil permeability (see Fig. 66)

L' is the factor of the slope length; a ratio which compares soil loss from
   the investigated plot and soil loss from the unit plot with a length of 22.13 m.
   It may be derived from the equation

$$L' = (\frac{L}{22.13})^p \tag{5.15}$$

where L is slope length measured from the water divide (m)
      p is an exponent with the value of 0.3 to 0.6; for slope gradients I≦10% the
      exponent is usually considered as being p = 0.5, for slope gradients
      I>10%, p = 0.6.
      S' is the factor of the slope gradient; a ratio which compares soil loss from
      the investigated plot and soil loss from the unit plot with a 9% gradient.
      It may be derived from the equation

$$S' = \frac{0.43 + 0.30I + 0.043I^2}{6.613} \qquad (5.16)$$

where I is slope gradient in %.

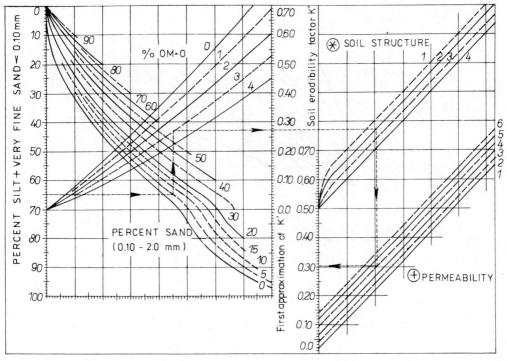

Fig. 66.   Nomogram for the determination of factor K'.

The authors of the equation (5.14) recommend that the factors of slope gradient and slope length be used in combination, i.e. as the combined L' S' factor for which they have recommended the following equation be applied:

$$L'S' = \frac{Lp}{100} (1.36 + 0.97I + 0.1385I^2) \qquad (5.17)$$

where L is the length of the plot measured from the water divide (m)
      I is the slope gradient (%).

The authors have constructed a nomogram (Fig. 67) for the quick determination of the L' S' factor.

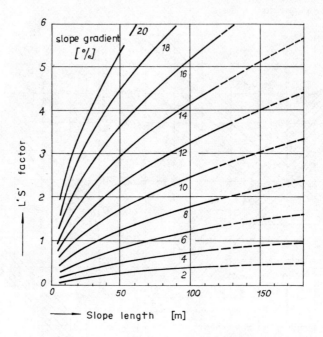

Fig. 67.   Nomogram for the determination of the L´ S´ factor.

C´ is the crop management factor – a ratio which compares soil loss from the
investigated plot to the soil loss from the unit plot provided the action of the
other factors is constant. In order to determine this factor in relation to the
changing protective effects of the vegetative cover and soil tillage in the course
of the year the authors divided the year into five crop stage periods:

- period 1: rough fallow, turn ploughing to seeding
- period 2: seeding-from seed-bed preparation to one month after planting
- period 3: establishment from 1 to 2 months after spring or summer
          seeding
- period 4: growing and maturing crop from end of period 3 to crop harvest
- period 5: residue or stubble – from harvest to ploughing or new seeding.

The authors compiled the table of C´ values from 10,000 data obtained at 47
monitoring stations (Table 5.3).
P´ is the factor expressing the effects of conservation – it is a ratio which
compares the soil loss from the investigated plot with soil loss from the standard
plot cultivated up and down the slope gradient. The values of the P´ factor for
contour farming are given in Table 5.4.

For contour strip cropping the authors recommend that the P´ value be reduced by
50%. Terracing is dependent on slope gradient and so the effective length of the
slope is reduced. The procedure is to use the same P´ value as for strip cropping
and the L´ S´ factor appropriate to the gradient and spacing between the terraces.

TABLE 5.3. Ratio of soil loss from crops to corresponding
            loss from continuous fallow – factor C´
            (Portion of 100–line published table;
            Wischmeier, 1960)

Cover, sequence, and management	Crop-stage Period (%)				
	1	2	3	4	5
1st year corn after meadow, RdL*	15	30	27	15	22
2nd year corn after meadow, RdL	32	51	41	22	26
2nd year corn after meadow, RdR**	60	65	51	24	65
3rd or more year corn, RdL	36	63	50	26	30
Small grain w/meadow seeding: (1) In disked corn residues					
After 1st-corn after meadow	–	30	18	3	2
After 2nd corn after meadow	–	40	24	5	3
(2) On disked corn stubble, RdR					
After 1st corn after meadow	–	50	40	5	3
After 2nd corn after meadow	–	80	50	7	3
Established grass and legume meadow	–	–	0.4	–	–

*RdL, crop residues left and incorporated by plowing
**RdR, crop residues removed

TABLE 5.4. P´values for contour farming

Slope gradient (%)	P´values for contour farming
1.1–2.0	0.6
2.1–7.0	0.5
7.1–12.0	0.6
12.1–18.0	0.8
18.1 and more	0.9

The given model is applicable for straight slopes. For slopes with irregular gradient G. H. Foster and W. H. Wischmeier[8] recommend that the slope be divided into segments with the same slope gradient and that the L' S' factor be calculated from the equation

$$L'S' = \frac{\sum\limits_{j=1}^{n} S_j \lambda_j^{1.5} - S_j \lambda_{j-1}^{1.5}}{\lambda_e / (22.13)^{0.5}} \tag{5.18}$$

where S_j is the factor of the slope gradient of the j-th segment of the slope calculated from equation

$$S_j = \frac{0.43 + 0.30I + 0.0431^2}{6.613} \tag{5.19}$$

where I is the slope gradient (%)
λ_e is the length of the investigated slope (m)
λ_j is the distance from the water divide to the bottom end of segment j (m)
n is the number of segments.

In order to speed up the calculation of the L' S' factor using equation (5.18) a nomogram has been constructed (Fig. 68) in which the equation applies

$$U_j = \frac{S_j \lambda_j^{1.5}}{22.13^{0.5}} \tag{5.20}$$

Factor L' S' for the whole slope will be obtained from the equation

$$L'S' = \frac{\sum\limits_{j=1}^{n} U_j - U_{j-1}}{\lambda_e} \tag{5.21}$$

A similar method may be used when the effects of varying soil properties in different parts of the slope, i.e. the variable K' factor, is considered for conversation practice.

The Wischmeier and Smith empirical model of soil erosion is based on a vast amount of data on erosion in various parts of the USA obtained over a 50-year period within a national soil conservation programme. It is widely applied not only in the USA but in a number of other countries where it has been tested and modified to suit local conditions.

Empirical models of soil erosion processes have the disadvantage of comprising complex interactions of many factors whose simplification may result in reducing the accuracy of calculations below acceptable limits. Another disadvantage is their limited local applicability, i.e. the model can only be applied to the locality for which the equations were derived or to localities with similar conditions. The use of the empirical model in different conditions will always require modifications considering the effects of local factors.

5.3 CONCEPTUAL MODELS OF EROSION

The recent development of computer technology has initiated efforts to express erosion processes using mathematical relations. Conceptual models have been built up allowing the observation of the individual stages of the erosion process, the flow of water and soil in the slope profiles, using mathematical relations. These models are being constructed on the basis of mathematical relations which apply in hydraulics, hydrology, etc.

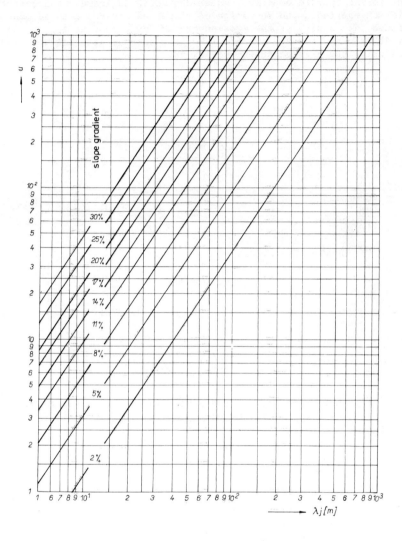

Fig. 68. Nomogram for the determination of the L´ S´ factor
for slopes with a variable gradient.

Conceptual models usually proceed from the assumption that the erosion process takes
place in inter-related stages. The following stages are the most important:
- the detachment of soil particles by rain
- the transportation of soil particles by rain
- the detachment of soil particles by surface runoff
- the transportation of soil particles by surface runoff.

Proceeding from the above determination of the stages of the erosion process L. D.
Meyer and W. H. Wischmeier[16] have built up a model which is a transition from
empirical to conceptual models. The dynamics of the four stages of the erosion
process are expressed by the basic equations. They then extended these fundamental

equations to incorporate empirical equations depicting namely the effects of soil properties on erosion. The investigated slope was divided into segments for which they defined the results of the erosion process by comparing the transportation capacity of rain and surface runoff with the amount of detached and transported soil particles using the diagram shown in Fig. 69.

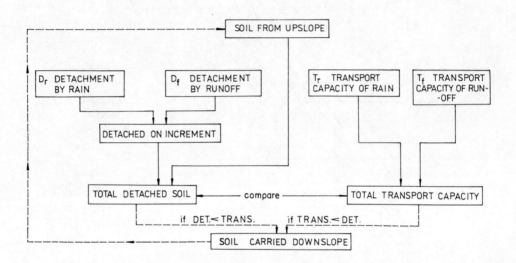

Fig. 69.   Model of erosion process after L. D. Meyer and
           W. H. Wischmeier.

L. D. Meyer and W. H. Wischmeier expressed the individual stages of the erosion process by the following equations:
    - the detachment of soil particles by rain written as

$$D_r = S_{DR}A_i i^2 \tag{5.22}$$

where D_r is the amount of soil particles detached by rain
      A_i is the area of the investigated section of the slope
      i rainfall intensity
      S_{DR} is the coefficient dependent on soil properties.

    - the transportation of soil particles by rain is expressed by the equation in the form

$$T_r = S_{TR}Ii \tag{5.23}$$

where T_r is the amount of soil particles removed by rain
      I is slope gradient
      S_{TR} is the coefficient dependent on soil properties

- the detachment of soil particles by surface runoff was expressed by the equation

$$D_F = S_{DF}A_iq^{2/3}I^{2/3} \qquad\qquad (5.24)$$

where D_F is the amount of soil particles detached by surface runoff
A_i is the area of the investigated section of the slope
q is the magnitude of surface runoff
S_{DF} is the coefficient dependent on soil properties.

- the transportation of soil particles by surface runoff was expressed by the equation

$$T_F = S_{TF}q^{5/3}I^{5/3} \qquad\qquad (5.25)$$

where T_F is the amount of soil particles transported by surface runoff
S_{TF} is the coefficient dependent on soil properties.

The application of this model requires that the coefficient of soil properties be known and correctly used and that other factors which affect soil cohesion and soil infiltration capability be considered.

G. R. Forster and L. D. Mayer[7] constructed a model in the form

$$D_F + R_{DT} = \frac{\partial G_F}{\partial x} \qquad\qquad (5.26)$$

$$\frac{D_F}{D_c} + \frac{G_F}{T_c} = 1 \qquad\qquad (5.27)$$

where D_F is the amount of soil particles detached by surface runoff
R_{DT} is the amount of soil particles detached by rain
G_F is the amount of soil particles transported by surface runoff
D_c the soil detachment capacity of surface runoff ($D_C = C_D\,x_*\,$ I)
T_c the transportation capacity of surface runoff ($T_C = C_T\,x_*\,$ I)
x_* xL_0^{-1} relative distance
x is the distance of the investigated profile from the water divide
L_o is slope length
I is slope gradient
C_D, C_T are coefficients of the effects of tangential stress, acceleration by gravity, roughness of soil surface, runoff.

From equations (5.26) and (5.27) the authors derived the resulting equation for soil loss on a straight slope in form

$$G_* = x_* - \frac{1}{\alpha}\,(1-\theta)(1-e^{-\alpha x_*}) \qquad\qquad (5.28)$$

where G_* is the relative soil loss on the slope in relation to the transportation capacity of runoff at the end of the slope $G_* = G_F T_{Uo}^{-1}$
θ is a parameter expressing the detachment of soil particles by rain $\theta = L_o R_{DT} T_{Uo}^{-1}$
α is the parameter expressing the detachment of soil particles by surface runoff $\alpha = L_o D_{Uo} T_{Uo}^{-1}$; index "o" designates values determined for the foot of the slope.

When deriving this equation the authors assumed that the detachment and transportation capacity of surface runoff is proportional to tangential stress to the power of two thirds and they assumed that Chézy's formula for steady state flow would apply for movement of soil particles by surface runoff.

The wide range of assumptions for deriving the equation and the large number of coefficients which have to be experimentally ascertained have prevented the wide application of this erosion model.

These shortcomings are removed by the conceptual model of erosion constructed by M. Holý and the team of researchers at the Institute of Irrigation and Drainage of the Faculty of Civil Engineering, Czech Technical University, Prague (J. Mls, J. Váška, J. Pretl and,K. Vrána) based on the principle of the conservation of mass and energy of surface runoff (see Chapter 4). Following the substitution of values obtained by routine local measurement the model may be used for calculating the surface runoff in any profile of the slope. The amount of washed soil particles may be found from the relation between surface runoff and soil wash derived from long years of research (Fig. 70a and b).

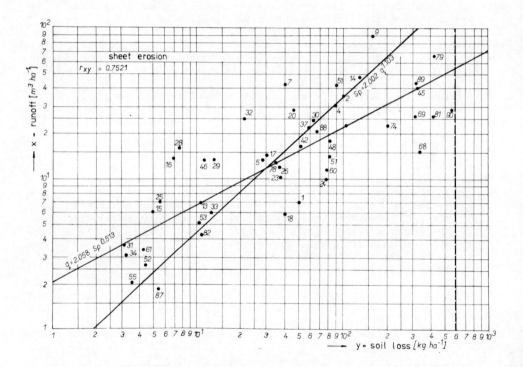

Fig. 70a.   Relation between surface runoff and soil loss due
            to sheet erosion.

5.4   PREDICTING EROSION INTENSITY IN THE CATCHMENT

In order to be able to predict the silting and pollution of water storage reservoirs and water courses it is necessary to know the assumed amount of sediments flowing not only from the slope but from the whole catchment. This question was dealt with by a number of authors but has hitherto not been satisfactorily resolved.

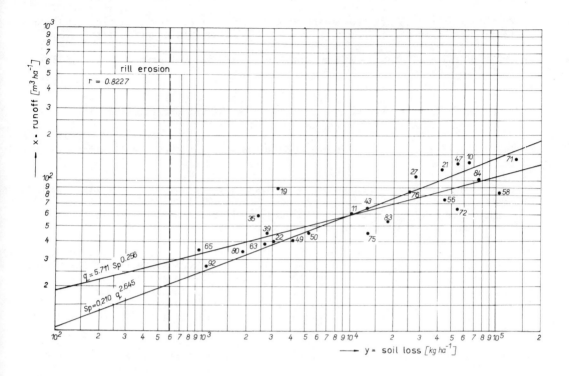

Fig. 70b. Relation between surface runoff and soil loss due
to rill erosion.

B. V. Polyakov[20] investigated the dependence of the sediment content in water
courses on runoff from the catchment, the gradient of slopes and a number of other
factors which he included in the so-called erosion coefficient which he expressed
as

$$G' = f_1(o, I, A)$$ (5.29)

where G' is the sediment content removed from the catchment by the water flow
o   is the annual amount of water flowing from the catchment
I   is the average gradient of the catchment
A   is the erosion coefficient expressing local conditions (type of soil,
cultivation, vegetative cover, etc.).

B. V. Polyakov proceeded from the assumption that the amount of sediments transported
from the catchment annually by water flow is in linear dependence on the annual
amount of water flowing in the water flow. He therefore determined the value of
coefficient A from the equation

$$A = \frac{\rho}{kI^n}$$ (5.30)

where $\rho = \dfrac{G'}{O} = f_2(I, A)$

B. V. Polyakov obtained coefficient A by direct field measurement and observation.

W. W. Anderson[1] presents the empirical equation derived from investigations made in 13 catchments in the forest regions of California. The equation is written

$$\log e = -2.044 + 0.866 \ \log q + 0.370 \ \log P - 1.236 \log C \qquad (5.31)$$

where e is the annual amount of sediments ($m^3 \ km^{-2}$)
     q is maximum annual flood flow ($m^3 \ s^{-1} \ km^{-2}$)
     P is the area of the main water course in the catchment ($m^2 \ km^{-2}$)
     C is the density of the vegetative cover (%).

Similar equations were derived for other catchments, but their application is limited to the local area.

An attempt at applying the universal soil-loss equation (5.14) for determining erosion intensity for the whole catchment was made by J. R. Williams and H. D. Berndt[25]. The individual members of the equation outside of the rainfall factor R´, were defined with regard to their application for the catchment area.

The soil erodibility factor K_p' (in this case the factor of catchment soil erodibility) was determined as being the weighted average of soil erodibility factor for different types of soil. The gradient of catchment I_p' is a weighted average from mean slopes between contours and the area adjacent to the contour. The path of surface runoff L_p' is determined as being 50% ratio of the total area of the catchment and of the total length of flows. Factors C_p' and P_p' are weighted average values C´ and P´ for the respective parts of the catchment area.

The results were tested by measurements of the amount of sediments flowing through the profile at the lower boundary of the catchment and calculated data were found to be in relatively good agreement with the measured values.

Equation (5.14) for calculating erosion intensity in the catchment area was also used by G. A. Onstad and G. R. Foster[19]. The rain factor R´ was substituted with the so-called energy factor W´ which they expressed as

$$W' = 0.5R_{st} + 4.33Q q_p^{\ 1/3} \qquad (5.32)$$

where R_{st} is the rainfall factor (in E´I´ units)
     Q   is the depth of runoff (cm)
     q_p maximum intensity of runoff (cm per hour)

The other factors of equation (5.14) are defined in the same manner as in the universal soil loss equation.

For calculating the intensity of the erosion process the authors divided the catchment into zones with approximately identical conditions, i.e. morphology, soil, vegetation, etc. Erosion intensity is defined for a unit of a section of each zone and the definition is then extended to the whole catchment using the appropriate transformation diagram. In the zones of the catchment area factors K´, S´, C´ and P´ of the universal soil loss equation (5.14) have constant value. In each zone the area is specified from which runoff flows directly into the water course and from these into the next zone. The intensity of sheet and rill erosion is determined for each zone, from which total erosion intensity can then be determined.

In Czechoslovakia the problem of predicting the amount of sediments in the catchment has been studied by O. Dub[4] who expressed the erosion activity of water in the catchment in dependence on surface runoff efficiency. The equation is written

$$G' = f(\bar{L}, C) \tag{5.33}$$

where G′ is the amount of sediments flowing from the catchment
      $\bar{L}$ is the efficiency of surface runoff
      C is the erosion coefficient expressing local conditions (type of soil, crop management methods, vegetative cover, etc.).

The map of contours and the map of isolines of average specific runoffs may be used to calculate the ideal efficiency of water n_i for an arbitrary specific runoff q applying the equation

$$n_i = \frac{1000qh}{102} \qquad [\text{kW km}^{-2}] \tag{5.34}$$

where q is specific runoff $(1 \text{ s}^{-1} \text{ km}^{-2})$
      h is the altitude of the measured point above the erosion base level, i.e., the water level at the outlet point.

Equations (5.33) and (5.34) may be used for deriving for the known area of catchment P the equation

$$\frac{G'}{P} = gf(n_iC) \tag{5.35}$$

where g is soil loss $(t \text{ km}^{-2})$
      n_i the specific efficiency of water (kW km^{-2}).

Erosion coefficient C may be determined for local conditions from the equation

$$C = \frac{g}{kn_i{}^\alpha} \tag{5.36}$$

where k is the coefficient of proportionality,
      $\alpha < 1$.

M. Holý[10] investigated the course of the silting of gravel collectors on stream tributaries of the Elbe river in the Bohemian Central Mts. (České Středohoří). The silting of one of the collectors built in 1908 on the Býnov brook, the sinistral tributary of the Elbe, is shown in Fig. 71.

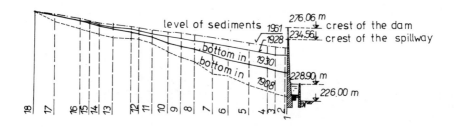

Fig. 71. Course of gravel collector clogging.

Archival data on the silting of several gravel collectors and field measurements
conducted in the area have allowed M. Holý to derive, using B. V. Polyakov's
equation[20], the equation for ascertaining the volume of silt from the catchment
in dependence on time. The equation is written

$$W = \frac{1}{\gamma} \; E'IqPt \tag{5.37}$$

where W is the volume of sediments (m^3)
     γ is the specific gravity of sediments (Nm^{-3})
     I average slope of the catchment area

$$I = \frac{H_A - H_B}{\sqrt{P}} \qquad \text{where}$$

     H_A is the altitude of the highest point in the catchment
     H_B is the altitude of the investigated profile
     P is the area of the catchment (km^2)
     q is specific runoff from the catchment $(m^3 \; s^{-1} \; km^{-2})$
     t is the duration of silting (s)
     E' is the erosion coefficient dependent on local conditions which may be
        obtained from the formulas applying for all other known members of the
        equation $(N \; m^{-3})$.

REFERENCES

1. Anderson, H. W., *Current Research on Sedimentation and Erosion in California
    Wildlands*, IASA, pp.173-182, 1962.
2. Bennet, H. H., *Elements of Soil Conservation*, New York, Toronto, London, 1955.
3. Cablík, J. and Jůva, K., *Soil Conservation*, SZN, Prague, 1963.
4. Dub, O., Erosion Intensity and its Determination by Hydrological Methods,
    In: *Vodohospodářský časopis*, SAV 1-2, 1955.
5. Dýrová, E., Permissible Slope Length Using Interception Ditches, *Final Report
    on Research Task A-0-29-23/4*, Brno.
6. Flaxmann, E. M., Predicting Sediment Yield in Western United States, *J. of
    Hydraul. Div. Am. Soc. Civ. Engrs.* Vol.98, H412, 1972.
7. Foster, G. R. and Meyer, L. D., *Closed-form Erosion Equation Sedimentation*,
    Colorado, 1972.
8. Foster, G. R. and Wischmeier, W. H., Evaluating Irregular Slopes for Soil
    Loss Prediction, *Transactions of the ASAE*, 2, 1974.
9. Frevert, R. K., *Soil and Water Conservation Engineering*, New York, 1955.
10. Holý, M. and Čáslavský, P., Contribution to the Problem of Reduction in
    Storage Capacity due to Silt Deposits under High-water Conditions,
    Transactions, V. Congress ICID, New Delhi, 1963.
11. Holý, M., *Soil and Water Conservation*, SNTL/ALFA, Prague, 1978.
12. Hudson, N., *Soil Conservation*, London, 1971.
13. Kohnke, H. and Bertran, A., *Soil Conservation*, New York,Toronto, London, 1959.
14. Kutílek, M., *Water Management in Soil Science*, SNTL/SVTL, Prague, 1978.
15. Levi, I. I., *Dinamika ruslovykh potokov*, Moskva-Leningrad, 1957.
16. Meyer, L. D. and Wischmeier, W. H., Mathematical Simulation of the Process of
    Soil Erosion by Water, *Transactions of the ASAE*, 1969.
17. Mirckhulava, G. E., *Inzhenernyie metody rascheta i prognoza vodnoy erozii*,
    Moskva, 1970.
18. Musgrave, G. W., Quantitative Evaluation of Factors in Water Erosion, *Journ.
    Soil and Water Conservation* 2, 1974.
19. Onstad, C. A. and Foster, G. P., Erosion Modeling on a Watershed, *Transactions
    of the ASAE*, Vol. 18.2, 1975.
20. Polyakov, B. V., *Gidrologicheskiy analiz i raschety*, Leningrad, 1946.
21. Polyakov, B. V., *Gidrologicheskie raschety pri proektirovanii sooruzheniy na
    rekach malych basseynov*, Moskva, 1948.

22. Shvebs, G. J., Empiricheskaya zavisimost dlya kolichestvennoy pover-khnostnogo smyva, *Sb.Rabot po gidrologii*, No.1, Gidrometeoizdat, 1959.
23. Váška, J., Modelling Water Erosion Processes. Thesis, Faculty of Civil Engineering, Czech Technical University, Prague, 1975.
24. Velikanov, M. A., *Rusloviy protses*, Fizmathiz, 1958.
25. Williams, J. R. and Berndt, H. D., Sediment Yield Computed with Universal Equation, *Journal of Hydraulic Division ASCE*, Vol.98, H 412, 1972.
26. Wischmeier, W. H., Smith, D. D. and Uhland, R. E., Evaluation of Factors in the Soil Loss Equation, *Agric. Eng.* 39, 1958.
27. Wischmeier, W. H. and Smith, D. D., Predicting Rainfall-Erosion Losses from Cropland East of the Rocky Mountains, *Agricultural Handbook*, 282, Agric. Research Service, US Department of Agriculture, 1965.
28. Wischmeier, W. H., Relation of Soil Erosion to Crop and Soil Management, *International Water Erosion Symposium, Proceedings. ICID*, Prague, 1970.
29. Zingg, A. W., Degree and Length of Land Slope as it affects Soil Loss in Runoff, *Agric. Eng.* 21, 1940.

6. Theory of Wind Erosion

The process of wind erosion may be divided into three stages, namely:
- the detachment of soil particles
- the transportation of soil particles
- the deposition of soil particles.

The first two stages are effected by the turbulent flow of ground wind with an energy that is capable of overcoming the gravitational force of soil particles, the third stage occurs when the energy of the wind drops below this limit.

6.1 EROSIVE ACTION OF WIND

The kinetic energy of the wind acts on the soil surface, it detaches soil particles and chemical substances which are bonded to them and puts them in motion. This action of the wind is known as deflation and when kinetic energy decreases deflation ceases and is followed by the process of deposition. It follows from the equation for the calculation of kinetic energy written

$$E = \frac{1}{2} mv^2 \qquad (6.1)$$

where E is kinetic energy (J)
     m is mass (kg)
     v is velocity ($m\ s^{-1}$).

that at equal mass of the air current the value of kinetic energy depends on the velocity of the flow. Wind velocity is next to pressure the main indicator of the force of wind. The degree of wind force on the Beaufort scale in dependence on its velocity and dynamic pressure are given in Table 6.1. Dynamic pressure is given at barometric pressure 101.32 kPa.

The pressure of wind against a flat surface, perpendicular to the direction of the flow may be expressed by the equation

$$p = \frac{a}{2g} v^2 \qquad (6.2)$$

where p is wind pressure (Pa)
     g acceleration due to gravity ($m\ s^{-2}$)
     v is wind velocity ($m\ s^{-1}$)
     a is the specific gravity of the air ($N\ m^{-3}$), whose value depends on
       temperature ($t°C$) and barometric pressure (kPa).

TABLE 6.1 Velocity and pressure of wind

Degree B.S.	velocity (mile per hour)	pressure (Pa)
0 - calm	1	0.0
1 - light air	1-3	0.29-1.96
2 - light breeze	4-7	1.96-7.85
3 - gentle breeze	8-12	7.85-19.61
4 - moderate breeze	13-18	19.61-39.23
5 - fresh breeze	19-24	39.23-68.65
6 - strong breeze	25-31	68.65-107.87
7 - moderate gale	32-38	107.87-166.71
8 - fresh gale	39-46	166.71-245.17
9 - strong gale	47-54	245.17-343.33
10 - whole gale	55-63	343.33-431.11
11 - storm	64-73	431.11-598.21
12 - hurricane	>73	>598.21

The dynamic pressure of storms reaches considerable values.

The most important factors for deflation are the direction of the wind and its velocity at ground level. The soil surface reduces wind velocity and this effect is most evident in hilly terrain, less so in flat countryside. Wind velocity increases with the height of the wind above the ground surface (Fig. 72).

The direction of the wind is given by the angle of the air current to the soil surface. Impact angle α may be used to express the so-called windwardness of surface k_n in the form

$$k_n = \cos\alpha \tag{6.3}$$

The most vulnerable areas to wind erosion are those where $k_n = 1$, i.e., where the direction of the air current is parallel to the soil surface; areas where $k_n = 0$, i.e., where the air current is perpendicular to the soil surface, are not susceptible to deflation.

Besides the velocity of dynamic pressure and the direction of the wind soil deflation also depends on the duration of wind action and on its frequency.

6.2 MOVEMENT OF SOIL PARTICLES BY THE ACTION OF WIND

The movement of soil particles is caused by the wind forces exerted against the soil surface. The average wind velocity near the ground increases exponentially with height above the ground surface. At a certain point near to the ground surface wind velocity is zero. The location of this point depends on the roughness of the soil surface and on the density and height of the vegetation. Above this velocity the average wind velocity will rapidly increase, then less rapidly as the height above the soil surface increases. In the zone above zero velocity, the wind is turbulent

and is characterized by eddy currents moving in all directions at variable velocities
(1,4). Turbulent wind puts the soil particles into motion.

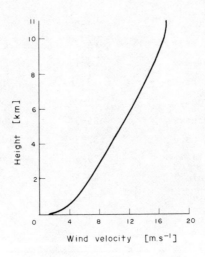

Fig. 72.   Relation between the wind velocity and the
           height above the ground surface.

There are three types of movement of soil particles depending on the size of soil
particles.

6.2.1 Movement of Soil Particles in the Form of Suspension

Very fine soil particles with a grain diameter <0.1 mm when put in motion move in
the form of an air suspension. Stoke's law which relates the velocity of falling
through a fluid to the square of its diameter applies to the movement of these
fine particles. The velocity of the fall of these fine particles is so small that
when they are lifted they remain suspended in air owing to turbulence and to the
eddy flow of wind for a very long time.

The average velocity of fine particles in calm air may be calculated after Stokes
(cit.5) and may be written

$$F = \frac{2}{g} \; \frac{r^2(\rho_1 - \rho)}{\mu} \; g \tag{6.4}$$

where F   is average velocity of soil particles movement (m s^{-1})
      g   is acceleration due to gravity (m s^{-2})
      μ is the coefficient of the viscosity of the medium (Pa s)
      ρ_1 is the density of the suspended particles (kg m^{-3})
      ρ is the density of the medium (kg m^{-3})
      r   is particle radius (m).

By substituting $\mu = \rho v$ where v is the kinematic coefficient of viscosity, equation (6.4) will be modified to the form

$$F = \frac{2}{g} \frac{r^2(\frac{\rho_1}{\rho} - 1)}{v} g \qquad [m.s^{-1}] \qquad (6.5)$$

The equation will apply for particles with a grain diameter <0.1 mm.

Dust storms carry vast amounts of fine suspended particles considerable distances.

6.2.2 *Movement of Soil Particles by Saltation*

This type of soil movement is by far the most important as it effects the transportation of the greatest amount of soil matter. The movement takes place in a series of low bounces over the surface and it occurs among medium sized particles which are light enough to be lifted off the surface but too large to go into suspension. The size of particles moved in this way is between 0.05-0.5 mm, most frequently between 0.1-0.15 mm.

The path of a particle moving by saltation after N. Hudson[3] is shown in Fig. 73.

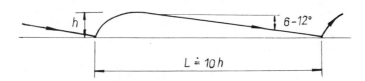

Fig. 73.   The path of a soil particle moving by saltation
          after N. Hudson.

The soil particle is lifted off the surface by the energy of the wind. The particle rises into the airstream and very shortly this vertical speed declines under the force of gravity. At the same time the particle picks up lateral velocity from the wind. Having risen to a peak which is only a few centimetres high it begins to fall but continues to accelerate laterally under the force of the wind and so returns to the soil in a long flat glide path striking the soil with great energy which puts further soil particles into motion.

6.2.3 *Movement of Soil Particles by Creep*

Movement of soil particles by creep is the movement of particles with a grain diameter from 0.5-1 to 2 mm which are pushed by the force of wind and by other particles moving with the wind. Theoretically there is no upper limit to the size of particles which may be rolled along the surface.

The speed at which the rolling of soil particles along the surface is effected may be calculated from the equation derived by M. A. Velikanov[6] and written

$$\frac{v^2}{gd} = \alpha + \frac{\beta}{d} \tag{6.6}$$

where v is the ground velocity of wind (m s^{-1})
      g is acceleration by gravity (m s^{-2})
      d is the soil particle diameter (m)
   α, β are empirically determined constants
      $\alpha = 14$, $\beta = 0.006$.

Equation (6.6) applies for soil particles between 0.1 to 5 mm in size.

By solving equation (6.6), Velikanov obtained the so-called rolling velocity of the wind which he termed critical for the respective soil particle sizes. Critical velocity may be calculated from the equation

$$v_{kr} = \sqrt{gd(\alpha + \frac{\beta}{d})} \qquad [m.s^{-1}] \tag{6.7}$$

Wind velocity is measured in meteorological stations usually at 8 m above ground surface. The relation between wind velocity at this height v´ and the ground velocity of the wind v may be written

$$v´ = 14.88v \qquad (m\ s^{-1}) \tag{6.8}$$

According to equation (6.7) the movement of particles with diameter d by particle creep occurs at a measured wind velocity at a height of 8 m above ground surface and may be expressed by the equation

$$v´ = 46.5\sqrt{14d + 0.006} \qquad (m\ s^{-1}) \tag{6.9}$$

The forms of soil particle movement after N. Hudson[3] are shown in Fig. 74.

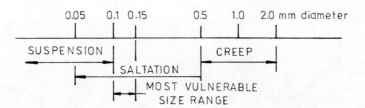

Fig. 74.   The forms of soil particle movement after
         N. Hudson.

6.2.4 *Distance to which Soil Particles are Transported by Wind Erosion*

The distance to which particles are transported by wind erosion depends on their size
and on the type of movement. Informative values are given in Table 6.2. The height
distribution of particles blown by wind is given in Table 6.3.

TABLE 6.2 The distance of transportation of soil particles
           by wind

Diameter of soil particles (mm)	Distance of transport
0-1	some meters only
1-0.125	1-1.5 km
0.125-0.0625	some kilometers
0.0625-0.0312	over 300 km
0.0312-0.0156	over 1500 km
below 0.0156	non limited

TABLE 6.3 Above ground height of soil particles transport-
           ed by wind

Height above soil surface (cm)	0-5	5-10	10-15	19-96
Amount of soil matter (%)	57.0	18.5	8.5	16.0

A typical example of the transportation and deposition of soil matter by wind erosion
are sand dunes in inland sand deserts and in coastal areas.

The smallest formation which results from the transportation of matter over small
distances are ripples with a height of 3-5 cm reaching a maximum of 10 cm and whose
width is approximately ten times their height. They are a typical phenomenon of
sand deserts.

When the wind current carrying the sand meets an obstacle, its velocity is reduced
and the sand is deposited, forming drifts with a mild windward slope (5-12°) and a
steep leeward slope (30-45°). The characteristic shape of a drift is shown in Figs.
75 and 76.

Drifts of regular shapes are only formed in areas with winds blowing prevalently
in one direction. Groups of such formations very often occur reaching heights of
10-20 m, in some cases up to 70 m. These drifts move with the wind and cover
villages, farmland, etc. They occur in the Karakum desert in Turkmenia (known as
barkhans), in the Sahara desert, and Gobi desert, etc.

In coastal areas dunes are formed, i.e., sand waves of approximately triangular cross
section stretching along the sea coast perpendicular to the direction of the wind
blowing in from the sea. Dunes occur in several rows and move inland. They reach
considerable heights piling to heights of up to 150 m on the African coast of the
Mediterranean. In some areas they dam river mouths which results in swamps being
formed on their windward slopes.

Fig. 75. Drifts in Sahara desert (photo by courtesy of
          ČTK, Prague).

Suspended soil particles travel farthest. Fine dust from the Sahara has been observed in northern Germany, Great Britain and in Scandinavia, at a distance of more than 3000 km from the Sahara desert, dust from Kansas was observed in New York at a distance of 2000 km from source, etc.

Dust from Africa has formed a 10 cm layer of silt in the coastal areas of Italy, southern France and the Tyrol over the past 300 years[2]. The layer of silt from fine dust blown by the wind in the catchment area of the Chuang-Che river reaches a thickness of 100 m.

Fig. 76. Drifts in Sahara desert (photo by courtesy of
ČTK, Prague).

REFERENCES

1. Beasley, R. P., *Erosion and Sediment Pollution Control*, Iowa, USA, 1972.
2. Cablík, J. and Juva, K., *Soil Conservation*, SZN, Prague, 1963.
3. Hudson, N., *Soil Conservation*, BT Batsford Limited, London, 1973.
4. Chepil, W. S., Dynamics of Wind Erosion, *Soil Science*, I-V, 1946.
5. Pasák, V., Wind Erosion on Soils, *Scientific Monographs*, VÚM Zbraslav, 3/1970.
6. Velikanov, M. A., *Gidrologiya sushi*, Leningrad, 1948.
7. Zingg, A. W. and Chepil, W. S., Aerodynamic of Wind Erosion, *Agric. Eng.* 31, 1950.
8. Zingg, A. W., Chepil, W. S. and Woodruff, N. P., Sediment Transportation Mechanics: Wind Erosion and Transportation, *J. Hydraul. Div., Proc. Am. Soc. Civ. Eng.*, Vol.91, 1965.

7. Intensity of Wind Erosion and Predicting Wind Erosion

Wind erosion intensity is determined similarly to water erosion, i.e., from soil loss from a unit area over a certain period of time. Like water erosion it is affected by a number of natural and anthropogenic agents. In determining the intensity of wind erosion several other important agents should be taken into consideration.

7.1 RELATION BETWEEN WIND EROSION INTENSITY AND EROSION AGENTS

The principle erosion agent effecting the initiation and course of wind erosion is the wind, other significant agents are the soil characteristics, the climate, the roughness of the soil surface, the vegetative cover and the length of the area along the direction of the wind.

As for climatic agents (Section 3.1) the most important are the wind, precipitation, temperature and evaporation. Humid soil is more stable than dry soil owing to the cohesion of soil particles. For this reason the soil moisture content is one of the main factors affecting the intensity of deflation.

The soil moisture content is determined by the amount and rate of precipitation and is affected by temperature, atmospheric moisture and wind which determine evapotranspiration and thereby also the loss of soil moisture. It is so significant that with the factor of wind velocity it is sometimes used to express soil erodibility[4].

This relation is expressed in the form

$$E_p = \frac{v^3}{w^2_{ef}} \qquad (7.1)$$

where E_p is soil erodibility
v is wind velocity
w_{ef} is effective soil moisture, i.e., the moisture between soil particles resisting tension induced by evaporation.

Effective soil moisture is dependent on precipitation and on temperature and the equation may be written

$$w_{ef} = \frac{S}{T^2} \qquad (7.2)$$

where S is precipitation over a given period of time
T is temperature over a given period of time.

134

The physical, chemical and biological state (Section 3.3) of the soil determines its resistance to the kinetic energy of the wind. Of special significance for the resistance of soils to the action of the wind is a stable structure, the size of soil particles and the soil moisture content.

The dividing line between erodible and nonerodible particles is not distinct and varies in dependence on wind velocity and on particles' size and weight. Most susceptible to erodibility are particles with a diameter of around 0.1 mm. Wind velocity needed for putting such particles into motion is about 16 km per hour at a height of 0.3 m above ground. Particles with a diameter <0.02 mm resist blowing by the direct action of wind on the smooth soil surface. Their cohesiveness hinders the blowing of larger particles mixed with them.

In case the soil surface composed of fine soil particles is roughened in such a manner that it is projected into the turbulent layer of the wind fine particles erode until the soil surface is levelled by the wind[2]. Wind velocity above the smooth and rough surface is given in Fig. 77.

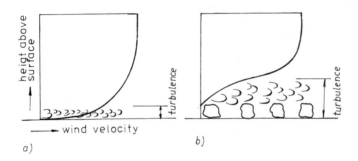

Fig. 77. Wind velocity above the smooth and rough surface
          after N. Hudson.

A roughened soil surface reduces wind velocity. Surface roughness reduces deflation but it also increases the turbulence of the wind thereby exposing soil surface to greater wind forces. Therefore, if the variations in the height of the roughened surface are too great, the benefits derived may be reduced. The optimal variations in surface roughness are 50-127 mm[1].

Vegetation reduces wind velocity at the surface and adsorbs much of the force exerted by the wind. Soil particles are trapped by the vegetation and it protects them from the direct impact of the wind. Dense and high vegetation is most effective and the residue of vegetation anchored in the soil is beneficial for reinforcing the soil profile and thereby for reducing wind erosion.

The increase in the rate of erosion with distance downwind over an unprotected eroding area has been described as avalanching. The longer the downwind area, the greater the number of particles dislodged in this manner. The soil particles accumulate on the soil surface until they are put in motion by a wind of sufficient force. The amount of moving erodible soil particles increases from zero on the downwind edge to a maximum that the wind can sustain. The value of the maximum of eroded particles depends on the length of the eroding area in the downwind direction.

This means that any obstructions that can break up the length of the eroding area
will reduce the intensity of deflation and that crop strips, shelterbelts, wind
barriers, etc. will accomplish this and are therefore justified.

7.2 DETERMINATION AND PREDICTION OF WIND EROSION

Scientists determining wind erosion intensity have attempted to qualify wind
erosion factors and to relate them in order to determine the intensity of erosion.
In view of the lack of adequate experimental data on wind erosion obtained in field
measurements these scientists, unlike those who constructed models of water erosion,
had to rely on data obtained by modelling deflation in experimental wind tunnels.
From data obtained in field conditions and in wind tunnels W. S. Chepil[3] obtained
the equation

$$E_p = f(P', C', K', L, V') \tag{7.3}$$

where E_p is soil loss owing to wind erosion (t ha^{-1} per year)
      P' is the factor of soil erodibility
      C' is the climatic factor related to wind velocity and soil moisture
      K' is the factor of soil roughness
      L is the length of the eroding area along the prevailing wind direction
      V' is the factor of vegetative cover.

The values of factors needed for the calculation of E_p have been studied by many
scientists in the US.

The erodibility factor P' expresses the potential soil loss (t ha^{-1} per year) from
an unprotected area with a bare flat soil surface without crust. Its value
increases with the percentage decrease of soil fraction with a grain diameter of
>0.84 mm. Factor P' obtained by investigations in the area surrounding Garden City
in Kansas, USA is given in Table 7.1.

TABLE 7.1 Factor of soil erodibility P'

Particles of dry soil >0.84 mm %	Units (t ha^{-1} per year)									
	0	1	2	3	4	5	6	7	8	9
0	–	694.9	560.4	493.2	437.1	403.5	381.1	358.7	336.2	313.8
10	300.4	293.7	286.9	280.2	271.2	263.3	253.3	244.3	237.6	228.6
20	219.7	213.0	206.2	201.7	197.3	192.8	186.1	181.6	177.1	170.4
30	165.9	161.4	159.2	154.7	150.2	145.7	141.2	139.0	134.5	130.0
40	125.5	121.0	116.6	114.3	112.1	107.6	105.4	100.9	96.4	91.9
50	85.2	80.7	74.0	69.5	65.0	60.5	56.0	53.8	51.6	49.3
60	47.1	44.8	42.6	40.3	38.1	35.9	35.9	33.6	31.4	29.1
70	26.9	24.7	22.4	17.9	15.7	13.4	9.0	6.7	6.7	4.5
80	4.5	–	–	–	–	–	–	–	–	–

Values given in Table 7.1 also apply for a slope where its length along the prevailing wind direction is more than 150 m.

K' is the factor of roughness of an unprotected soil surface. It may be determined
from variations in height in the microrelief of the soil surface (Fig. 78).

The climatic factor C´ has been determined from the relation between the annual average wind velocity, modified with regard to standard height of measurement (30 ft - approximately 9 m), effective soil moisture expressed by the Thornwaite moisture index P-E (cit. 8) and the amount of transported soil in the form

$$C´ = 34.483 \ \frac{v^3}{(P - E)^2} \tag{7.4}$$

where v is the average wind velocity at 10 m above the soil surface (mile per hour)

P-E is the precipitation- evapotranspiration difference.

The values of the C´ factor for Kansas and parts of Nebraska, Colorado, Oklahoma, New Mexico and Texas are given in Fig. 79.

L is the length of the unprotected area along the prevailing wind direction.

The factor of vegetative cover V´ was investigated for different conditions from the type and amount of vegetation in a given area.

An example of factor V´- the vegetative cover provided by stubble, is given in Fig. 80.

The general functional dependence of the potential soil loss at wind erosion E_p is shown by equation 7.3. The mathematical relation between the given factors is complicated owing to various interactions and the equation can therefore not be solved multiplying together the individual factors as has been the case in water erosion.

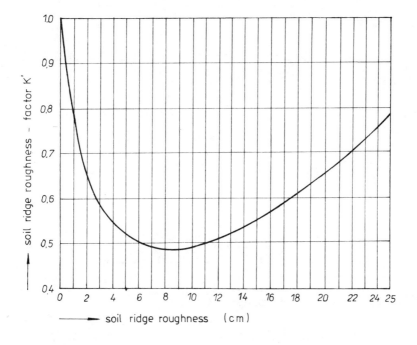

Fig. 78.   The determination of K´ factor.

Formulas have only been derived giving the dependence of E_p on individual factors. The equation may then be solved in five steps where each step evaluates the effects on E_p of one variable.

1st step – the soil loss $E_1 = P'$ is determined. This loss occurred on an
          unprotected plot with a bare soil surface at a measured proportional
          content of aggregates with particle diameter >0.84 mm
2nd step – the effects are found of the roughness of the soil surface $E_2 = P' \cdot K'$
3rd step – the effect of wind velocity and soil moisture is introduced
          $E_3 = P' \cdot K' \cdot C'$
4th step – the factor of the length of the plot is introduced $E_4 = P' \cdot K' \cdot C' \cdot f(L)$.

In this case the simple multiplication of L is not possible because the values L, P', K', C', V' are inter-related. The graphic solution of the 4th step is shown in Fig. 81.

5th step – the vegetative cover is introduced
          $E_5 = E_p = P' \cdot K' \cdot C' \cdot f(L) \cdot f(V')$

In this case in view of the inter-relations between V' and the other factors a simple multiplication is not possible. The authors therefore recommend the solution shown graphically in Fig. 82.

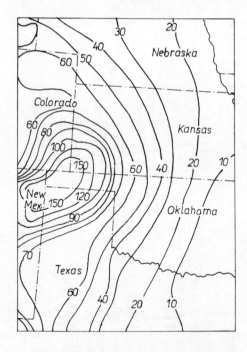

Fig. 79.   The determination of C' factor.

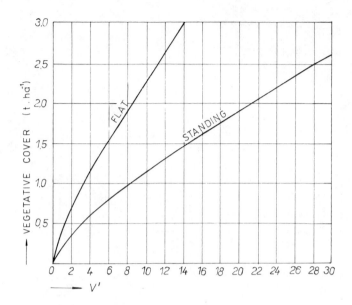

Fig. 80.   The determination of V´ factor.

The solution of equation (7.3) is relatively difficult. E. I. Skidmore, P. S. Fischer and N. P. Woodruff[6] therefore developed computer solutions written in FORTRAN IV. Tables and nomograms are stored in the computer memory which from input data on conditions characterizing a given area will determine E_p.

The application of equation 7.3 for different conditions would require that a modification be made to the given factors and possibly that new factors be derived. This was done for Czechoslovak conditions by K. Vrána[7].

Following long-term observations of wind erosion in field conditions and in an experimental wind tunnel (Fig. 83), V. Pasák[5] derived an equation for the determination of wind erosion intensity in the form

$$E_p = 22.02 - 0.72 \ P'' - 1.69V + 2.64v \qquad\qquad (7.5)$$

where E_p is soil erodibility during wind action in time t = 15 min (g m^{-2})
     P'' is the content of nonerodible particles in the soil   (>0.8 mm) (%)
     V   is relative soil moisture and is determined by the relation of instantaneous moisture corresponding to the wilting point
     v   is wind velocity at ground surface level (5 cm above soil surface) (m s^{-1}).

The equation is solved using a nomogram (Fig. 84).

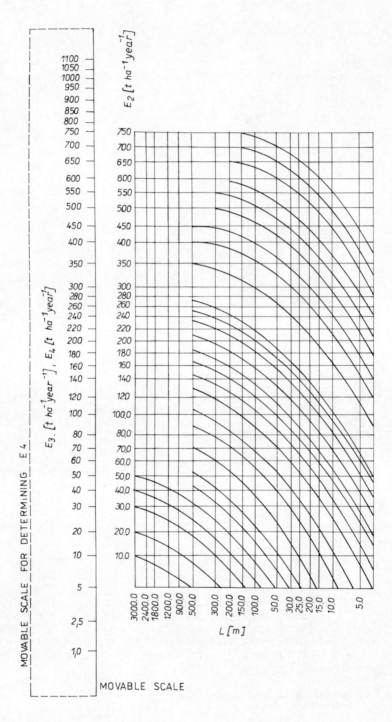

Fig. 81. The graphic solution of the 4th step of the
          equation (7.3).

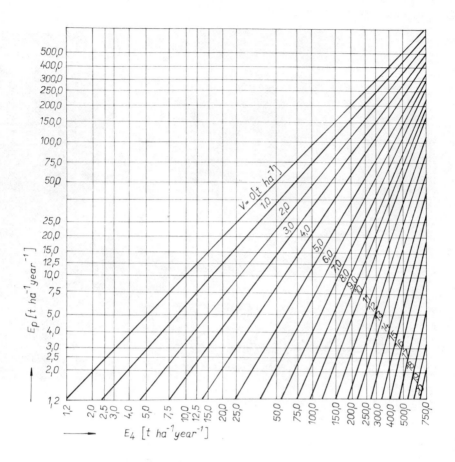

Fig. 82. The graphic solution of the 5th step of the
equation (7.3).

For practical application the nomogram includes not only the above ground wind
velocity but also the velocity of the wind at 8 m above the soil surface. Soil
erodibility E_p is determined by soil loss in kg ha^{-1} by the so-called erodibility
rate I_e, i.e., the relation of deflation and its permissible value. The permissible
value of deflation (I_e = 1) was determined by V. Pasák as being the average removal
of soil particles from a soil having a 60% proportion of nonerodible particles.
V. Pasák recommends that the proportion of non-erodible soil particles in soil be
determined by running an average sample of soil from the surface layer which has
been dried in the air through a sieve with a 0.8 mm mesh.

Fig. 83. Experimental wind tunnel after V. Pasák.

The equation is written

$$P'' = \frac{p}{c} \ (100) \tag{7.6}$$

where p" is the content of non-erodible soil particles in the soil (%)
     p   is the weight of the sample after sifting (g)
     c   is the weight of the sample before sifting (g).

On the basis of experiments conducted in a wind tunnel V. Pasák determined the
informative values of the wind for different soil types. This was done by convert-
ing the wind velocity in the tunnel to wind velocity measured in meteorological
monitoring stations (8 m above ground level). The values are given in Table 7.2.

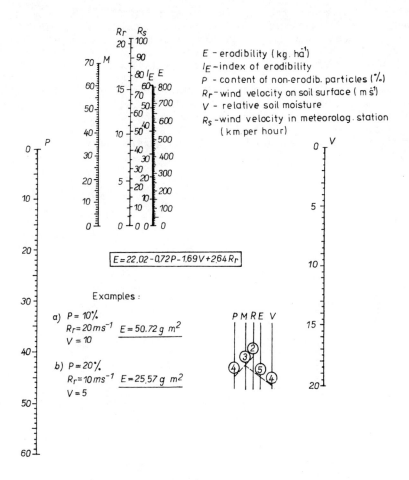

Fig. 84. The nomogram for solving the equation (7.5).

TABLE 7.2 Extreme wind velocities (km per hour) for
different soil types

| | Soil | |
Type	Dry	Moist
Sandy soil	16	29
Loam sandy soil	16	82
Sand loamy soil	31	58
Loamy soil	106	106

REFERENCES

1. Beasley, R. R., *Erosion and Sediment Pollution Control*, Iowa, USA, 1972.
2. Hudson, N., *Soil Conservation*, B.T. Batsford Limited, London, 1973.
3. Chepil, W. S. and Woodruff, N. P., The Physics of Wind Erosion and its
    Control, *Advances in Agronomy* 15, 1963.
4. Chepil, W. S., Influence of Moisture on Erodibility of Soil by Wind, *Proc. Soil
    Sci. Am.*, Vol.20, 1956.
5. Pasák, V., Wind Erosion on Soils, VÚM, Zbraslav, *Scientific Monographs* 3, 1973.
6. Skidmore, E. L., Fisher, P. S. and Woodruff, N. P., Wind Erosion Equation:
    Computer Solution and Application, *Proc. Soil Sci. Am.*, Vol.34, No.6, 1970.
7. Vrána, K., *Determination of Wind Erosion in Czechoslovakia*, Prague, 1977.
8. Woodruff, N. P. and Siddoway, F. H., A Wind Erosion Equation, *Proc. Soil Sci.
    Am.*, Vol.29, No.5, 1965.

8. Erosion and Environmental Control

When evaluating the unfavourable effects of erosion processes emphasis has usually
been placed on their influence on changes occurring in soil fertility which in turn
reduce agricultural production. The significant development of anthropogenic
erosion has turned the attention to other consequences of erosion, namely to
environmental pollution by excessive silting and to the pollution of water resources
by chemicals removed by surface runoff from farmland in catchment areas. These effects
of soil erosion are extremely unfavourable because erosion functions as a nonpoint
pollution source over large areas where large-scale environmental control installations have to be built. As compared with point pollution, e.g., waste effluents from
industries and agriculture which may be disposed by suitable equipment, erosion as a
nonpoint source is an important factor in the pollution of water resources.

8.1 POLLUTION SOURCES

The rapidly growing economic activity of society, namely the significant intensification of agricultural production, is accompanied by the increased application of
mineral fertilizers and pesticides as well as by the growing quantities and expanding range of waste products. These substances are removed by erosion and pollute
surface and groundwater resources.

8.1.1 Mineral Fertilizers

The endeavour to increase agricultural production both in the advanced and in the
developing countries is generally evident by the growing application of mineral
fertilizers, namely N, P, K. Data published regularly by the UN Food and Agriculture
Organization show the rapid trend of growth of the consumption of these fertilizers
in all continents.

The rapid growth in the application of mineral fertilizers arises from the need to
increase the soil nutrient content which is exhausted by the intensive exploitation
of the soil. This rise is bound to continue in developing countries, especially
with the introduction of large-area irrigation schemes. Advanced countries with a
highly intensified chemicalized agriculture will have to consider the consistent
implementation of comprehensive measures to raise soil fertility and the possibility
of applying mineral fertilizers as only one component of comprehensive soil treatment.
The UN Economic Commission for Europe pointed out as early as in 1973[3] that high

fertilizer rates applied to agricultural land in advanced European countries should not be substantially increased. E. C. Bui of the USA[3] justified the 50% increase in fertilizer application in the USA over the past 50 years by saying that the costs of fertilizers were lower than the costs of any other means of increasing farm production.

Agriculture has gradually become the main source of nitrogen and phosphorus in water resources. This is evident from data published by McCarty[11] (Table 8.1).

TABLE 8.1   N a P content in different sources (after McCarty).

Source	Nitrogen (mil. kg/year)	Phosphorus (mil. kg/year)
Domestic wastes	500 to 720	90 to 225
Industrial wastes	> 450	x
Runoff from agricultural land	680 to 6800	54 to 545
Runoff from non-agricultural land	180 to 860	68 to 340
Wastes from live-stock production	> 450	x
Runoff from urban areas	50 to 500	5 to 77
Atmospheric precipitation	13 to 265	1.5 to 4

x - no data

Owing to the intensification of agriculture in Central Europe a significant increase of nitrogen, phosphorus and potassium in the soil has been recorded. Systematic observations will have to be conducted to assess the relation between the intensification of agriculture and the pollution of water resources. In Czechoslovakia, for example, it is estimated that 0.5 kg P_2O_5 ha^{-1} per year are released into the ground water by seepage, and 20% of the applied phosphorus fertilizers are washed into surface flows. Data obtained from measurements made by the Czechoslovak Academy of Sciences[14] (Fig. 85) document the growing level of surface water pollution in dependence on increased rates of mineral fertilizers. Increased fertilizer rates have a closer relation to water pollution in the Vltava river than to the increase in the yields of the principal crops in the area. This indicates that it would be sensible to combine the application of mineral fertilizers with other measures aimed at raising soil fertility, e.g., improved technology, irrigation, etc., and that fertilizer rates should only be increased after all other techniques having no adverse environmental impacts had been tried.

Water pollution is specially unfavourable in reservoirs serving for water supply. In countries where ground water resources make up only a small proportion of the water supply and where the major part of the population will have to be supplied with potable water from surface water resources, the pollution of water by chemicals released by erosion from agricultural land creates a serious problem which is difficult to solve.

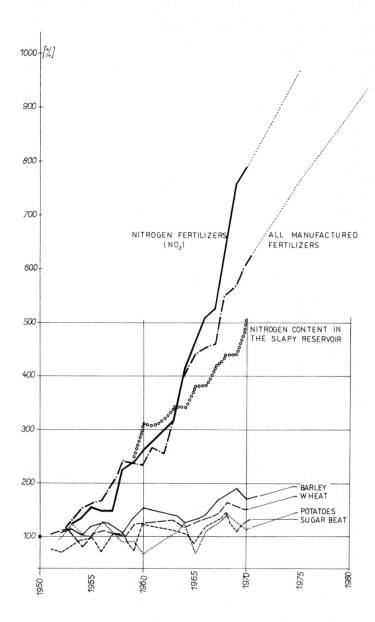

Fig. 85. Water pollution due to fertilizers.

8.1.2 Pesticides

Pesticides-chemical substances used in agriculture to protect the vegetation from
pests, are a significant source of soil and surface and ground water pollution.
Certain types of pesticides are harmful to man into whose body they are transported

through the food chain cycle[6]. Another unfavourable aspect is the persistence
with which some of the residual pesticides are retained in the soil from which they
are removed by erosion processes and released into surface and ground waters. I. G.
Nash and E. A. Woolson[13] state that twenty years after application they found that
up to 40% of some of the applied pesticides were still present in the soil. As
compared with easily degradable chemical substances, persistent pesticides are
transported by moving water and may become the source of pollution for soil and water
at a great distance from the place of applications.

According to their purpose pesticides may be classified as:

- herbicides
- fungicides
- insecticides
- acaricides (for the control of mites and ticks)
- molluscocides (for snail and clam control)
- rodenticides (for rodent control)
- repellents (substances used to repel animal pests).

Herbicides, weed-killers, are either selective, i.e., they destroy only certain types
of weeds without damaging cultural plants, or non-selective which destroy all treated
plants.

The predecessors of today's herbicides were copper salts and arsenical compounds which
are highly persistent and are retained in the soil for many years. Cases have been
known of fish poisoning by copper salts washed into water reservoirs by surface runof
from orchards and vineyards[7]. Since 1930 Dinitro compounds have been used as herbi
cides, e.g., DNOC and Dinoseb which are dangerous for mammals but they break down
rapidly in the soil. These substances were replaced by a group of less toxic selecti
herbicides, such as MCPA and 2-4-D. The latest generation are the synthetic organic
group of non-selective herbicides of which Simazin and Monuron are highly persistent
Dalapon breaks down after 6-8 weeks and is used for the control of weeds along strear
and rivers without apparent ill effects. Paraquat is neutralized almost immediately
on contact with the soil.

Fungicides destroy fungi and moulds. Almost all of them are persistent. Copper
sulphates which in higher concentrations are dangerous for fishlife are often used fo
controlling blight on fruit trees and on potatoes. Fungicide dressings applied to
seeds are often based on organo-mercury compounds which are extremely persistent.
Dangerously high levels of mercury have recently been found in tuna fish in many
parts of the world and it is evident that such dressings are the original source of
the mercury which like DDT is absorbed by living tissues and can be concentrated by
the food chain from small amounts to highly dangerous levels.

Insecticides are effective substances for insect and animal pest control. Insecticides are classified as stomach insecticides, contact insecticides and fumigants
which are applied to the surface of the plant and systemic poisons which the plant
absorbs into its metabolism.

In 1945 a group of organo-phosphorus insecticides were produced of which some were
highly poisonous but broke down quickly in the soil, i.e., Parathion, Malathion,
Diazinon and systemics, such as Phosphamidon, Dimethoat, Di-chlorvos, etc. Insecticides based on organochlorines or chlorinated hydrobarbons are highly persistent,
e.g., DDT which is an efficient insecticide and is mainly used for pest control in
grain growing regions, namely in developing countries. In view of its toxicity,
high persistence and its retention and accumulation by living bodies, it is a
dangerous insecticide whose use has been prohibited in many countries. Other insecticides, such as Aldrin, Dieldrin, Heptachlor, Endrin and others are persistent and
are dangerous to mammals and birds.

The toxicity of pesticides is shown by their chemical structure - in a large group
of bonds the toxicity of heavy metals is in organo-metal bonds. In the process of
the breakdown of these bonds the heavy metals are often released and are involved
in further chemical reactions resulting in further toxic compounds. A classical
example of such a process are organo-mercury bonds. After the breakdown of the
organo-mercury bonds mercury and inorganic salts may be transformed by anaerobic
microorganisms into highly poisonous dimethylmercury which is soluble in fat where
it accumulates. In the years 1953-1960 many people died in Japan after having eaten
fish whose bodies contained dimethylmercury[6].

The problems of the effective control of pests, diseases and weeds which threaten
crops on a world scale is extremely urgent and chemicals used for this purpose are
as yet irreplaceable. The worldwide use of herbicides is expected to increase by
250% by 1990 as compared with 1975. The most important herbicides will also in the
future be triazines, derivatives of phenyl urea, phenoxocompounds, bonzoates and
carbamates. Only arsenite consumption is expected to decrease. The most important
insecticides whose consumption is expected to double by 1990 as compared with 1975
are organo-phosphates and carbamates while the use of chlorinated hydrocarbons will
considerably decrease. The use of physical insecticides which kill pests by their
physical properties will increase more than fivefold.

Dithiocarbamates will continue to be the most important fungicides. The use of
organic copper compound based fungicides will increase only very little and will
completely be discontinued before 1990. New types of pesticides are therefore sought
which would act selectively. The possibility of substituting chemicals with biological control is also being studied.

8.1.3 Wastes and Sludges

The total amount and range of wastes increase with the development of industrial
production, urbanization and intensification of agriculture. By their properties
wastes are classified into solid, liquid and gaseous. Their quantity and quality
differ from country to country depending on the level of civilization and on the
types and abundance of natural resources. The total annual production of solid
wastes in the USA amounts to 3.5 thousand million tons, of this 1.6 thousand million
tons are agricultural wastes, 1.1 thousand tons are industrial wastes and 360 million
tons are household wastes[6]. It is estimated that by the year 1980 some of these
values will double. The daily production of domestic waste is highest in the USA
where it amounts to 2-3 kg per inhabitant per day, in the FRG 1-2 kg per inhabitant
per day, in Czechoslovakia 0.55 kg per inhabitant per day. Waste production is
showing a constantly rising trend.

The amount and composition of industrial wastes depends on the extent of production
and on the quality and type of raw materials processed in production as well as on
the technology applied in the production process. The major part of industrial
wastes are inorganic substances, such as slag, dross, ash and debris which are
usually transported hydraulically to dumps, tips and sludge tanks. Organic wastes,
namely from sugar mills, dairies and distilleries are used in agriculture as fertilizers and feeds. Dangerous wastes are toxic, acid, alkali, putrifying and radio-
active wastes, less dangerous or harmless are inert and non-inert wastes and wastes
resembling solid household wastes. Agricultural wastes have become a grave problem
owing to the industrialization of agricultural production and the establishment of
large-scale livestock breeding farms where livestock is concentrated in one area and
the natural recirculation of animal and plant wastes has been eliminated.

The concentration of livestock breeding on farms with more than 10 thousand heads
of cattle or pigs no longer allows the use of all produced waste as manure. With

regard to biodegradable organic substances a large scale pig breeding farm with more than 20 thousand heads of pigs equals the pollution produced by a town with a population of 40 thousand inhabitants, a poultry farm with 150 thousand poultry equals pollution produced by a town with a population of 75 thousand inhabitants. L. Kaminsky[9] sets the average population equivalent value of pigs at 3.77 (excreta 0.64, urine 3.13), the total population equivalent of mature cattle is 33.1 (excreta 4.4, urine 28.7).

Solid urban wastes include solid wastes, street refuse high-volume wastes, low-volume industrial wastes and sludges from wastewater treatment plants. The amount of urban wastes varies from 100-250 kg per inhabitant per annum.

There are basically four types of waste and sludge disposal:

  - storage (tipping, dumping)
  - incineration
  - composting
  - re-processing, recycling using various industrial technologies.

The dumping and tipping of wastes and sludges and their composting and incineration have to be considered with regard to erosion processes. Tips, dumps and composts take up a considerable area which is exposed to erosion. Owing to atmospheric precipitation various substances are leached from tips and immature compost heaps and are removed to water flows and reservoirs, with the water they infiltrate into the soil and are transported into surface and ground waters which they pollute. Currently sludges and wastes are often mixed into the soil, to prevent water pollution. However, owing to erosion, various substances are released and removed by surface runoff and infiltrated water.

Small particles, especially from incineration, are windblown and in areas with intensive wind erosion they are a dangerous source of atmospheric pollution.

8.2 TRANSPORTATION OF POLLUTANTS

The particles and substances are removed from the soil by the energy of raindrops, by the shear stress of surface runoff or by wind and are transported by runoff into the hydrographic system or carried by infiltrating water into the deeper soil horizons, or transported by wind.

In agricultural areas where mineral fertilizers have been applied the soil removed by erosion processes always has a higher nutrient content than the parent material.. This is due to the higher nutrient content in the topsoil, fine soil particles of this layer with a large efficient surface which adsorbs chemical substances are more easily washed away. The ratio of the nutrient content in the removed soil to nutrients in the parent material is called the enrichment ratio (Table 8.2).

The movement of nutrients on slopes has been studied by M. Holý[5]. On all observed slopes he found a decreasing nutrient content in the top soil layer. The example in the decrease in organic carbon is shown in Fig. 86.

By a more detailed investigation of the curves of nutrient loss on erosion affected slopes M. Holý arrived at the conclusion that the curves may be replaced by second degree parabolas written

$$I_e = f\ (2p) \hspace{6cm} (8.1)$$

where I_e is the erosion intensity
2p is that parameter of the parabola which represents the course of organic
carbon content on the slope affected by erosion. He derived the
following expression for the parameter of the parabolic curve.

$$2p = \frac{x_2 \, x_3 \, (x_3 - x_2)}{x_3 \, (y_1 - y_2) - x_2 \, (y_1 - y_3)} \qquad\qquad (8.2)$$

where x_1, x_2, x_3 is the distance from the top of the slope ($x_1 = 0$)
y_1, y_2, y_3 is the per cent content of organic carbon in the soil at
distances x_1, x_2, x_3, from the top of the slope.

Values 2p may be used to determine the intensity of the loss of plant nutrients by
erosion on the affected slopes.

TABLE 8.2   Enrichment ratio for certain substances after
N. Hudson and D. C. Jackson

Plant nutrients	Extreme values		Normal values	
	low	high	low	high
Nitrogen	1.35	4.20	1.90	2.30
Organic carbon	1.35	4.20	1.80	2.20
Phosphorus	1.15	5.56	2.20	2.60

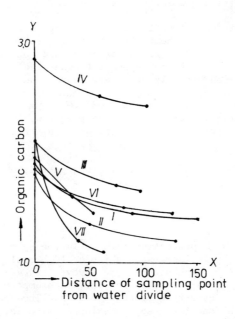

Fig. 86.   Decrease in organic carbon on slopes.

Pesticide residues are usually not readily soluble and are adsorbed on the surface
of fine soil particles and are carried in suspension. On the basis of a study of
the movement of pesticides from experimental plots C. T. Haan[4] states that more
than twice as much of the pesticide residues are carried by the removed soil as by
the runoff water.

8.3   CONSEQUENCES OF WATER POLLUTION BY EROSION

The judgment of the importance of water pollution by mineral fertilizers removed by
erosion depends on the use of the water. The gravest problem is posed by the
pollution of domestic water supplies. This is because some substances, namely
phosphates, nitrates and chlorides directly act on the human organism and may cause
direct health hazards. Nitrates can for instance, produce a condition in infants
known as methemoglobinemia. There do not exist any economical and efficient techniques for removing phosphates, nitrates and chlorides. For irrigation water, a
reasonable quantity of nitrogen and phosphate in the water can be an advantage.
Industry, apart from the food industry, is not likely to be adversely affected by
water with the usual concentrations of mineral fertilizers.

The most frequently occurring problem arising from pollution by mineral nutrients
is the effect on biological balance in streams, rivers, and lakes with the resulting eutrophication thereof. Eutrophication is the term given to the acceleration
of the normal biological process owing to the increased inflow of nutrients. This
causes the spectacular growth of algae, which gives an undesirable taste and odour
to water, may reduce the efficiency of filters by clogging and reduce available
oxygen to such an extent that fish and other aquatic life is affected. Eutrophic
water flows and reservoirs are unsuitable for recreation.

The application of pesticides in agricultural production is the principal source
of water pollution with various chemicals, very often toxic. This process is
accelerated by erosion. So far there exists no method for their efficient removal.
Pesticides render an undesirable and unpleasant odour and taste to water and they
are a health hazard for man not only by direct contact but through the food chain.
K. Mellanby[12] shows this by the example of a lake in the USA polluted by DDD, a
chemical similar to DDT, in a concentration of 0.015 ppm. In the food chain K.
Mellanby found a concentration of 5 ppm in the plankton, 10 ppm in small fish, 100
ppm in big fish and 1600 ppm in birds feeding on fish (this level was a health
hazard to the birds).

Some pesticides are retained by the soil for long periods of time and contaminate
the soil profile.

8.4   WATER AND SOIL POLLUTION CONTROL

Theoretically water and soil protection from pollution by mineral fertilizers is
possible by determining fertilizer rates which could be fully utilized by the
vegetation. In practice some substances are always leached into the soil because
they are consumed only gradually and during their consumption atmospheric precipitation with leaching and transporting capacity occurs. The only practical solution
is to reduce the surface runoff thereby lowering erosion intensity. Pollution
control may be achieved by prohibiting the use of highly toxic and highly resistant
substances as has been done with DDT in the USA and in most European countries. In
the continents of Asia and Africa DDT is still widely applied. The most effective
measure to prevent the spread of these substances outside the area of their
application is effective soil conservation. It should, however, be taken into
account that surface water percolating the soil carries dissolved substances into

the deeper layers of the soil profile. This is, however, less dangerous than their
direct transportation into the hydrographic system.

Waste tips and dumps must be controlled, carefully sited in such a manner as to
prevent the direct pollution of ground waters by leached substances. The dumped
wastes should be covered with a layer of compacted earth 15-20 cm thick. The last
layer of waste should be covered by a layer of earth 1-1.5 m thick. Dumps which
might pollute ground waters should be provided with a watertight lining; usually a
clay membrane.

REFERENCES

1. Biswas, A. K., *Water Development, Supply and Management*, Vol.2, United Nations
      Water Conference, Ottawa, Canada, 1978.
2. Dvořák, P., Handová, Zd., Holý, M., Koníček, Zd.; Kutílek, M., Říha, J. and
      Vrána, K., Activity of the ICID in the Field of the Human Environment,
      ICID Technical Memoirs No.2, New Delhi, 1974.
3. Economic Commission for Europe, *Proceedings, Seminar on Water Pollution by
      Agriculture and Forestry*, Vienna, 1973.
4. Haan, C. T., Movement of Pesticides by Run-off and Erosion, *Agricult. Eng.*
      52,1,25, 1971.
5. Holý, M., Specified Classification of Sheet Erosion, *Collection of Scientific
      Works*, Faculty of Civil Engineering, Prague, 1958.
6. Holý, M., Říha, J. and Sládek, J., *J. Society and the Environment*, Prague, 1975.
7. Hudson, N., *Soil Conservation*, BT Batsford, London, 1973.
8. Hudson, N. and Jackson, D. C., Results achieved in the Measurement of Erosion
      and Run-off in Southern Rhodesia, *Proceedings 3rd Inter-African Soil Conference*,
      Dalaba, 1959.
9. Kaminský, L., Testing of the Magnitude of Pollution Caused by Certain Types of
      Agricultural Wastes, *Vodní hospodářství* No.8/1974, Series B, pp.217-221.
10. Koníček, Zd., Origination of Sludges and Wastes and their Characteristics,
      Proceedings, ČSVTS, Kalová problematika 1975, České Budějovice.
11. McCarty, P. L., Sources of Nitrogen and Phosphorus in Water Supplies, *Journ. of
      American Water Works Association*, 59, 1967.
12. Mellanby, K., *Pesticides and Pollution*, Collins, London, 1967.
13. Nash, R. G. and Woolson, E. A., Persistence of Chlorinated Hydrocarbon Insect-
      icides in Soil, *Science* 157, 924, 1967.
14. Svatoš, J., *Principles of the Organized Development of the Intensification of
      Large-scale Agricultural Production as Limited by Water Resources*, Materials
      of the Commission for Water Management, ČSAV, Prague, 1975.
15. Wright, A., World Pesticides Worth $ 14 Billion by 1990, *Chemical Age* 112, 1976.

9. Erosion Control

Erosion control is indispensable in view of the expanding economic activity of society and the endeavour to use natural resources purposefully and economically. The objective of erosion control is to protect the two most valuable natural resources, i.e., water and soil and to prevent the occurrence of the unfavourable consequences which their deterioration could have for various branches of the national economy, namely for agriculture and water management and for the human environment.

The basic demand placed on erosion control is that it should be comprehensive. Any evaluation of water erosion processes and any erosion control scheme should conceive the catchment as being the basic unit in which an organic system of measures may suitably modify runoff conditions. This also applies for wind erosion which may partly be eliminated either by the vegetative cover or by control of soil moisture in the catchment.

Erosion control measures must be harmonized with the requirements placed by agricul-tural production, water management, transport, industry and other branches of the national economy in order to attain maximum economic effect while securing soil and water resources conservation. All activities in the catchment must be based on a scientific approach and a thorough knowledge of all existing and expected relations.

Erosion control measures cover three basic groups:

- agriculture and forestry
- technical control of sheet surface runoff
- technical control of concentrated surface runoff.

Special attention should be devoted to the prevention of gully and ravine erosion and to torrent control

9.1 THEORETICAL ANALYSIS OF PERMISSIBLE SLOPE LENGTH

Erosion control measures are aimed at the two fundamental erosion factors, namely slope gradient and slope length. In order to reduce their adverse effect on the origination and course of erosion processes it would be necessary to limit tangentia stress and the velocity of surface runoff to a value at which intensive disturbance of the soil surface does not occur. Terracing is the most effective measure for reducing slope gradient.

This technique is relatively costly and may usually only be applied to valuable cultures on good soils. Soil conservation practice is therefore usually oriented towards reducing unfavourable slope length. The so-called critical slope length is determined as the length on which sheet runoff becomes concentrated runoff and sheet erosion becomes rill and gully erosion. Erosion intensity on a slope which does not exceed critical length is not considered as being dangerous for agricultural production.

Various authors have studied critical slope length which they investigated theoretically and practically. As an example let us give the results obtained by J. Cablík[5] who investigated critical slope length for loamy soils on the basis of tangential stress and the velocity of surface runoff.

He expressed tangential stress T_k by the equation

$$T_k = \gamma_v y I \qquad\qquad [N.m^{-2}] \qquad\qquad (9.1)$$

where y is the depth of surface runoff (m);
         its expression is given by equation (4.9) in section (4.1)
      I is the slope of the area
      γ_v is the specific weight of water; γ_v = 9806   N m^{-3}.

Having obtained the permissible value of tangential stress for loamy soils after G. Strele[25] T_k = 10.79 N m^{-2}, using coefficient m = 24.8 (cit.5) and runoff coefficient o = 1 he obtained the critical slope length written

$$L_1 = \frac{m' T_k^2}{10^6\ oi}\ \frac{1}{I\ 3/2} \qquad\qquad [m] \qquad\qquad (9.2)$$

Applying equation 4.10 from section 4.1 for critical velocity and permissible velocity for loamy soils after M. A. Velikanov V_k = 0.24m s^{-1}[20] J. Cablík obtained the following equation for the critical slope length

$$L_2 = \frac{v_k^2}{m'oi}\ \frac{1}{\sqrt{I}} \qquad\qquad [m] \qquad\qquad (9.3)$$

For the calculation of critical slope length this expression applies which gives a lower value. It is derived from the equation

$$L_1 \gtreqless L_2 \qquad\qquad\qquad\qquad (9.4)$$

$$\frac{v_k^2}{m'oi}\ \frac{1}{\sqrt{I}} \gtreqless \frac{m' T_k^2}{10^6\ oi}\ \frac{1}{I\ 3/2} \qquad\qquad (9.5)$$

$$I \gtreqless \left(\frac{m' T_k}{1000 v_k}\right)^2 \qquad\qquad (9.6)$$

Where the upper sign applies critical slope length is determined from tangential stress, when the bottom sign applies slope length is determined from velocity.

The relation of the slope gradient and critical slope length using values T_k = 10.79 N m^{-2}, v_k = 0.24 m s^{-1}, m´ = 24.8, o = 1, and the mean intensity of rainfall i = 0.58 mm min^{-1} and duration t = 45-60 mins is shown in Fig. 87. With a different runoff coefficient the resulting values of the critical slope length must be multiplied by its reverse value.

M. Holy[11] introduced physical conditions for surface runoff from the slope for calculating the critical slope length and obtained the following equation:

$$L = am' \frac{T_k^2 B}{\gamma_v(aA\gamma_v 1 - T_k)} \quad I^{-1/2} \qquad [m] \qquad (9.7)$$

where m´ is the coefficient of soil roughness
T_k is permissible tangential stress (N m^{-2})
γ_v is the specific gravity of water (N m^{-3})
I is the relative slope of the area
A,B are values obtained from the substitute curve of specific runoff
(see sections 3.1.4.1)   A (m)   B (ℓ)
a is the coefficient dependent on soil properties.

M. Holý obtained this coefficient in field investigations from characteristic runoff intensity which he expresses as the difference between runoff intensities 10 mins after the start of rainfall for a period of 100 mins of rainfall duration divided by a period of 90 mins. The values of the coefficient for various slope gradients and for moderately heavy soils as well as the characteristics of runoff intensity 1.222 and 1.908 may be obtained from Fig. 88.

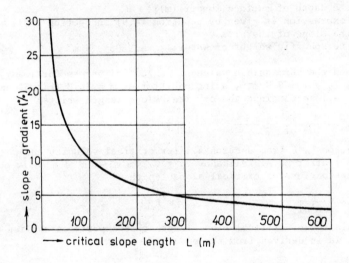

Fig. 87.   Critical slope length after J. Cablík.

The permissible slope length is still under study and new ideas about its calculation are still appearing.

9.2   AGRICULTURAL AND FORESTRY MEASURES

Agricultural and forestry techniques used for erosion control exist in·the correct location of cultures, a well designed layout of plots and communication system, correct cultivation of field and forest soils and the use of the protective effects of the vegetative cover.

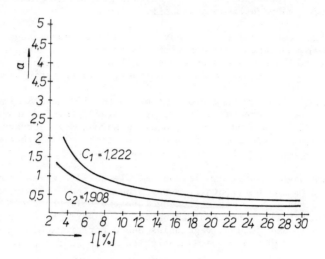

Fig. 88.   Values of the coefficient C.

9.2.1 Location of Cultures

The location of cultures has immense influence on the origination and course of
surface runoff and on soil erosion resistance. Different cultures provide different
conditions for the infiltration of precipitation into the soil and for the course of
surface runoff, they bind the soil by their root systems, they enrich it with organic
residues thereby upgrading its physical, chemical and biological properties, they
shade the soil and prevent the occurrence of excessive evaporation, they affect wind
flow in the ground layer of the atmosphere, etc. In an area susceptible to erosion
the location of cultures must therefore be totally subordinated to the demands of
erosion control. This is affected by the relief of the area. Water divide areas,
slopes and valleys have different hydrological and soil properties which determine
the suitability of cultures and the choice of cultivation techniques used. The
planting of cultures (field, meadows and pastures, vineyards, groves, etc.) is
conditioned by site conditions of which the most important are the climate and the
soil properties.

Water divides which cover the highest altitude areas are characterized by rough grained
permeable soils which easily absorb precipitation which infiltrates into the
deeper layers and flows out as subsurface runoff. The water divide is therefore
a suitable site for deep rooting cultures, such as forests and groves which by
turning surface runoff into subsurface runoff provide effective erosion control.
Afforested water divides, with good precipitation conditions are the main source
of subsurface water at the foot of slopes and in the valleys.

Soil permeability usually decreases downhill and surface runoff occurs on larger
collecting surfaces initiating erosion. In view of the fact that next to
precipitation the other important erosion determinant is slope gradient, cultures
planted on slopes should be chosen appropriately, Slopes with a gradient of more
than 36%, at erosion exposure of more than 20% should be afforested, slopes with a

gradient of more than 21% on slopes which are less exposed to erosion should be
planted to a permanent grass cover. A well kept forest and grass cover allow the
infiltration of precipitation into the soil and protect it from the destructive
effects of raindrops and of surface runoff.

It is usually not economical to afforest or to plant grass on the upper parts of
slopes with their gradients in areas with a mild relief and deeper soil layers. Such
areas may effectively be used for permanent cultures, such as vineyards and groves,
provided efficient soil conservation techniques, like terracing are applied.

The lower parts of slopes with a gradient of up to 21% and a maximum of 31% and
running into the valley are suitable for arable land provided appropriate agrotechnological, biological and in some cases technical soil conservation practices
are applied to protect the fields from water erosion. It is recommended that groves
and vineyards be planted on favourably exposed slopes.

Valley locations have heavier and less permeable soils, enriched by erosion processes
with fine particles and nutrients from the slopes. They are supplied with moisture
from surface and subsurface water flowing into the valleys, often from ground water
with a table near to the surface of the area. The soils have a high capillarity
and are a good site for fodder, vegetables and other crops highly demanding on moisture; sites with a high ground water table are suitable for permanent pastures.

The appropriate location of cultures with regard to the relief of the terrain is
shown in Fig. 89.

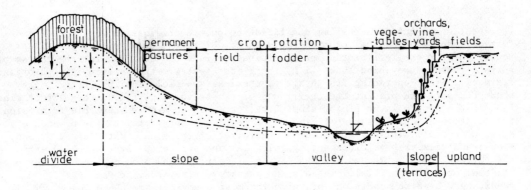

Fig. 89. Location of cultures with regard to the relief of
the terrain.

9.2.2 Shape, Area and Location of Agricultural Plots

In advanced countries the shape, area and location of agricultural plots is determined by the use of large-scale technology and mechanization. Best suited for this
purpose are continuous unbroken regular land units with identical slope and soil
conditions.

The best shape of plots is a rectangle with interior angles of $50°$-$60°$, cultivated along the longer side. The suitable relation of the lengths of the sides of the rectangle is 1:2 to 1:3 and 1:6 at the most. The length of the plot in an area not susceptible to erosion is given by the efficient use of mechanization.

The choice of the shape and area of plots in areas susceptible to erosion should aim to secure soil conservation. The shape and area of plots must be purposefully adjusted to the relief of the terrain which significantly affects the water and wind conditions in the area. The width of plots should never exceed the permissible limit (see section 9.1).

In areas which are susceptible to erosion, on slopes with a slope gradient of more than 5% land units have to be located with the longer side along the contour and contour cropped. Contour cultivation enhances the infiltration of precipitation into the soil and reduces the danger of surface runoff which initiates erosion. Significantly higher soil loss in downslope location and cultivation have in many cases underlined the necessity of locating fields along the contour in erosion sus- ceptible areas. One such finding (cit.5) proved that soil wash in the spring thaw from a plot located and cultivated downslope was 388.4 t ha^{-1} and surface runoff reached an average depth of 1.2 mm while on a contour located and cultivated plot under the same conditions soil wash was only 13.3 t ha^{-1} and the average depth of surface runoff 0.1 mm.

Stripping is extremely effective if crops with low erosion control efficiency alter- nate with crops that provide effective erosion protection to the soil. On the slopes high boundaries and terraces should be preserved which adjust the slope of the area as well as low boundaries which act as infiltration strips.

These measures should be taken in order to use the soil economically. The shape, area and location of plots should be selected in such a manner as to provide maximum erosion control allowing the use of the soil and the use of economically feasible agrotechnology and mechanization. Should this prove impossible other forms of land use should be considered.

9.2.3 Communications

A well established system of field paths which link the farm's production centre with the fields may become a good and worthy component of complex erosion control provided these communications are suitably located. Field paths dissect slopes thereby intercepting surface runoff; when a cross section and gradient of the path- ditches comply with the flow rate of water running from the slopes they divert the runoff into the recipient. These ditches are designed and built within the frame- work of other projects and systems for regulating and routing the water flowing from the area, i.e., infiltration strips, infiltration ditches, outlets, etc.

The main principle which should be observed in routing roads and field paths in areas susceptible to erosion is that they should be located on the ridge or as near to it as possible. The path has a small collecting area and does not require drainage, the retained water will be diffused into the surrounding terrain. It does not require any installations and maintenance is easier. Paths and other communica- tions which cannot be located near the ridge should have a mild slope approximately along the contour. A drainage ditch should be built at their upper end. Paths with a slope of 0.2-1% are suitable for the transportation and diversion of water in open canals, a slope of 1-5% does not pose any difficulties for transportation but requires special techniques to be applied for diverting the precipitation water retained in the ditches. Least suitable are paths and roads built transversely to the slope at a gradient of more than 5%. In such a case it is better to change

their direction several times or to use a combination of greater length in a mild
slope and smaller length in a steeper slope (Fig. 90).

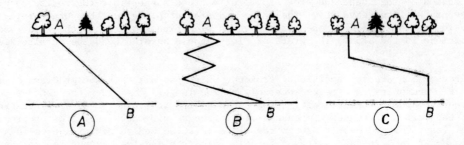

Fig. 90. Agricultural communications.

A system of peripheral communications running along the field contours is best for
erosion control (Fig. 91).

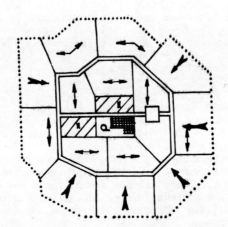

Fig. 91. Peripheral agricultural communications.

A three dimensional model of the area obtained from a stereoscopic image of aerial
photographs is a good aid for laying out the communication system in hilly terrain.

9.2.4 Soil Cultivation

The objective of soil cultivation in areas susceptible to soil erosion is to create favourable conditions for optimal crop harvests, to increase the soil's resistance to water and wind erosion, to enhance the infiltration of water into the soil, to create conditions for non-erosive runoff of water from the surface of the area and to safeguard the soil moisture supply.

Most resistant to erosion is soil with good physical, chemical and biological properties as in a crumb structure. Such soils have high cohesiveness and water permeability. For this reason the cultivation of soil in areas susceptible to soil erosion is guided by an attempt to form and preserve the crumb structure of the soil. Very effective in this respect is the formation of a sorption saturated complex in the soil which enhances the presence of organic substances supplied by fertilization, especially by lime fertilizers, composts with a suitable composition, etc.

Soil is prepared by mechanical treatment for planting cultural plants. For soil conservation purposes such mechanical treatment must not disturb the soil structure. It should enhance the infiltration of water into the soil and thereby contribute towards maintaining a favourable level of soil moisture. The number of cultivation operations which destroy the crumb structure of the soil should be minimal. S. K. Kondrashov[16] has described his observations made in the Taskhent area of Uzbekistan where the frequently loosened topsoil of the cotton plantations contained a mere 14% of crumbs while uncultivated land had a 65% crumb content and fields planted with lucerne a 94% content of crumbs.

In areas which are susceptible to water and wind erosion the agrotechnology of minimum cultivation has lately been introduced, i.e., farm machines are equipped with attachments and implements in such combination that cultivation and seeding is carried out in one operation. The soil surface between the rows remains rough and retains its crumb structure with considerable porosity which allows for a high accumulation of water and for its infiltration into the soil. To this tillage technique may be added fertilization and chemical protection which requires efficient haulage machinery but does not repeatedly disturb the soil by a series of subsequent operations.

Various types of soil conservation agrotechnologies are currently in the stage of research, development and trial.

Soils in areas susceptible to soil erosion should be contour cultivated. Cultivation may be slightly sloped towards the edges of the plots. Contour tillage and subsequent soil treatment and seeding bring about the interception of water running off from the slope surface in furrows and rows, its accumulation and diffusion over the soil surface and increased infiltration of runoff into the soil. Examples of the beneficial effects of contour cultivation on erosion intensity are shown in Table 9.1.

Contour tillage protects the soil from deflation. The furrow ridges form obstacles slowing down the velocity and force of ground winds. Soil particles blown from the furrow ridges are mostly deposited in the neighbouring furrows and they are not lost from the plot.

Contour tillage is also beneficial for uniform distribution of snow in the fields. Here it prevents the occurrence of hoarfrost and during the spring thaw it secures an even distribution to soil of moisture from melting snow.

TABLE 9.1   Relation of cropping direction and runoff

Tillage technique	Area of field (m²)	Precipitation (mm)	Runoff (mm)	Runoff coefficient
Downslope	2.000	33	1.2	0.040
Contour	3.000	31	0.1	0.003
Contour	10.000	56	0.0	0.000

Contour cultivation is suitable for areas where tractors and other machines are used. Contour cultivation is more efficient and is fuel saving as compared with downslope cultivation.

Special tractors are built for the contour cultivation of slopes with a gradient of more than 17%. These tractors do not tilt or deviate owing to slope. Implements are attached in such a manner that they do not slip downslope in turning and are controlled from the tractor.

Furrowing is recommended on slopes with a gradient such that erosion cannot be prevented by contour cultivation. This technique consists in ploughing a deeper furrow with a specially modified plough blade. The furrow is broken with soil at every 50-200 m so that it intercepts and retains water and snow as well as soil particles blown from neighbouring unprotected plots.

Good cultivation of the soil includes supplying moisture to the soil which increases its cohesiveness and thereby its resistance to the erosive action of water and wind. A favourable soil moisture content may be obtained by creating conditions for the infiltration of precipitation. Soil with favourable physical, chemical and biological properties intercepts water from liquid and solid precipitation and retains it in the soil profile

Water supplied to the soil by snow precipitation is very important. The uniform distribution of the snow cover may be enhanced by ploughing rough furrows and contour furrows before the winter season and by planting vegetation strips, using snow fences (Fig. 92) and making small snow ridges, etc.

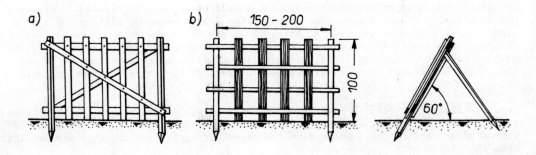

Fig. 92.   Snow fences.

The snow thaw is favourably affected by dark material, such as ash and peat, which is deposited on the snow in parallel strips and which by its colouring enhances the snow thaw. As a result, parallel strips are formed on the slope in which water intensively infiltrates into the soil profile.

In areas with a lack or shortage of precipitation, soil moisture should be supplied by irrigation.(12)

9.2.5 Use of Vegetation for Soil Conservation

Vegetation protects the soil from the effects of water and wind erosion. Very effective in this respect are forest and agricultural plants whose erosion control efficiency depends on the type of plants and on the conditions in which they are grown. Erosion control should make use of such properties of the vegetation which do not deteriorate soil quality but which conserve or possibly even improve it and which increase soil fertility. Vegetation may be used for conservation by:

   - soil conservation crop rotations
   - strip cropping
   - grassland farming
   - protective forest belts
   - afforestation.

9.2.5.1 Soil Conservation Crop Rotation

Soil conservation crop rotation is the distribution of agricultural plants into fields in such a manner as to secure the regular rotation of crops. Cereals, potatoes and beet, forage crops, hemp and flax are planted in rotation in order to preserve soil fertility and to secure high yields with regard to the previous crop.

A correct crop rotation is a significant erosion control measure provided the composition of the rotating plants is such as to contain as many erosion control plants as possible, e.g., perennial grasses, namely alpha-alpha.

The erosion control effect of the crop rotation is in direct proportion to the amount of perennial fodders. The exposure of the soil to erosion is increased by potatoes and sugar beet, etc. which do not provide sufficient cover to the soil. Table 9.2 shows soil wash under different cultures after N. I. Sus(26).

TABLE 9.2   Comparison of soil loss in different crops after N. I. Sus

Crop	Relative soil loss (%)
Forage mixture	0
Clover	1
Winter crops	50
Spring crops and fallow	100
Potatoes, sugar beet	200

In areas which are greatly exposed to erosion, soil conservation crop rotation is applied in combination with permanent meadows which are planted in horizontal strips on the arable land according to the principles of strip cropping.

Very important for wind erosion control are crop rotations in which perennial
plants are prevalent, and meadows rotate only with plants with a high above-ground
vegetal part and a dense undergrowth.

9.2.5.2 Strip Cropping

Strip cropping makes use of the erosion control effects of vegetation and its
favourable effect on water infiltration into the soil. It consists of alternating
strips of plants which do not sufficiently protect the soil from erosion, i.e. protected strips (potatoes, cereals) and protective strips (grasses) which protect the
strip of crop plants below the grass strip.

According to the type of erosion which is to be controlled, protective strips are
classified into:

  - contour strips which provide water erosion control
  - plant strips protecting the soil from wind erosion.

Water erosion control crop strips must be arranged in such a manner as to intercept
rain water running off from strips of crops with insufficient erosion resistance
and to ensure that this water infiltrates into the soil. Strip cropping, in which
two strips of potatoes, sugar beet or any other culture providing poor erosion control
or crops which have the same planting and harvesting date, should certainly be
avoided.

Protective strips are part of the erosion control crop rotation.

The width of the protected strips should not exceed the critical length of the
slope. The width of the protective strips is determined by their purpose. i.e.,
that the water running off from the protected strip which lies above the protective
strip should be intercepted and should infiltrate into the soil and that no rain
water which falls on the protective strip should flow away.

The width of the protective strip D may be expressed as

$$Dw = q_1 + q_2 \tag{9.8}$$

where D is the width of the protective strip (m)
      q_1 is precipitation pertaining to the protective strip $(m^3 . s^{-1})$
      q_2 is water running off from the protected strip $(m^3 . s^{-1})$
      w is the infiltration rate of the protective strip $(m\ s^{-1})$.

The diagram of the calculation is given in Fig. 93.

The equation may be written

$$q_1 = Di \qquad\qquad [m^3.s^{-1}] \tag{9.9}$$

$$q_2 = v_x y \qquad\qquad [m^3.s^{-1}] \tag{9.10}$$

where i is the intensity of calculated rainfall (see section 3.1.1) $(m.s^{-1})$
      v_x is the velocity of surface runoff $(m\ s^{-1})$
         (see section 4.1)
      y is the thickness of the layer of the runoff water at the upper end of
         the protective strip (m).

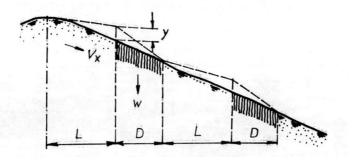

Fig. 93. Diagram for calculation of protective belt width

The expression (4.4) and (4.9) in section 4.1 therefore apply after substitution for v_x and y

$$Dw = Di + \sqrt{\alpha o i x} \cdot \sqrt{o i \frac{x}{\alpha}} \qquad (9.11)$$

where w is the infiltration rate of the protective strip (m s^{-1})
      o is the coefficient of runoff.

For the width of the protected strip x = L the equation for the width of protective strip D and protected strip L is

$$\frac{D}{L} = \frac{oi}{w-i} \qquad (9.12)$$

Infiltration rate w depends on the vegetation, soil properties and tillage methods applied. The numerical values of infiltration intensity vary depending on local conditions.

In conditions in Czechoslovakia, tilled soil without vegetative cover and soil planted with potatoes or sugar beet gives w = 0.2-0.3 mm min^{-1}, soil with a grass cover w = 1.3 mm min^{-1} etc.

Strips are contour planted (Fig. 94). A regular configuration of the terrain permits strips of uniform width and thereby makes an efficient and economical use of mechanization. An irregular configuration of the terrain makes their width variable depending upon changes in slope gradient.

In areas affected by intensive erosion permanent infiltration strips (buffer strips) are recommended (which are not part of the crop rotation) unless technical erosion control measures may be taken.

Wind erosion control strips with high crops (maize, sunflower, sorghum) alternate with strips of erosion conductive row crops, such as potatoes and sugar beet. The favourable effects of strip cropping will increase if the vegetation above-ground is left in the field as long as possible (Fig. 95).

Fig. 94.   Strip cropping in Iowa (photo by courtesy of the
           Soil Conservation Service, USA).

Wind strip cropping as a wind erosion control cropping procedure is the planting of regular farm crops in straight parallel strips at right angles to the direction of the prevailing winds.

9.2.5.3 Grassland farming for erosion control

Soils threatened by erosion which can neither be tilled nor usefully afforested should be turned into permanent grassland. Permanent grassland is established on erosion affected land in areas with an irregular configuration, quicksands, barren land, industrial tips, dumps, etc.

Erosion control can only be provided by a high-quality grass cover. Grassland planted in mountainous areas with light skeletal soils and ground water at greater depths under the surface of the terrain is usually poor and unable to protect the soil from erosion. Appropriate cultivation methods such as the retention of winter precipitation, fertilization, seeding of resistant grasses, etc., should therefore be applied.

Fig. 95. Wind erosion control strips (photo by courtesy of the Soil Conservation Service, USA).

In pastureland, overgrazing must be prevented because it ruins the turf which makes an effective cover for the soil.

Effective protection of the soil can only be provided by pastures with a well rooted growth and a prevalence of plant species resistant to grazing and puddling. Overgrazing and the heavy trailing of land by cattle and sheep may become the cause of gully erosion. This may be prevented by dividing large areas of pastureland into smaller units by fencing.

9.2.5.4 Protective forest belts

Protective forest belts are planted in areas threatened by erosion and not suitable
for crop planting. They should be planted in such length and width as to control
erosion in the whole cultivated area. Protective forest belts may be classified
into:

- windbreaks
- infiltration and retention forest belts
- shelter belts.

Protective forest belts are not only important for erosion control but also for
raising soil fertility. They do, however, take up part of the land which in areas
with a lack of agricultural land is an adverse factor. Their planting should therefore be carefully considered with regard to the area concerned.

9.2.5.4.1 Windbreaks. Windbreaks are mostly planted in flatlands exposed to violent
dust winds which carry away soil particles, cause snow drifts and reduce soil
moisture.

Windbreaks slow down and deflect the wind near the ground. The air current is forced
out of the microclimate and the resulting turbulence reaches the boundary layer of
the atmosphere by dynamically conditioned currents (Fig. 96). Some authors believe
[1,2,3,15] that these effects of air turbulence may cause local vertical precipitation. Windbreaks reduce intensive airflow at ground level throughout the protected
area and also the intensity of the airflow and thereby:

- check topsoil blowing
- check snow blowing in the winter season which allows a more even distribution
  of the snow cover and gives protection to the seeded crop plants
- reduce moisture evaporation and transpiration, thus making the
  soil resistant to erosion
- allow the infiltration of snow melt into the soil because of the lower intensity of snow thaw in protected areas.

The difference in the temperature of the ground layer of the atmosphere in the
windbreak area and that of the exposed area results in the formation of surface and
soil dew which with the transpiration of moisture from the trees increases the air
humidity.

The positive impacts of windbreaks increase with their area and with the humidity
of the climate. A. M. Alpatyev[2] states that ideally they should cover 5-8% of the
area.

By their density windbreaks may be classified as impermeable, permeable and semi-
permeable.

Impermeable windbreaks have a dense closed canopy and thick brush undergrowth. Their
disadvantage is the accumulation of snow inside the belt and in the summer a considerable increase in temperature on the leeward side. Impermeable windbreaks may be
used to retain snow along communications, to reduce noise pollution and to filter
soild air contaminants.

Semi-permeable windbreaks have a 20% permeable loosely thatched canopy. According
to Soviet experience such windbreaks are most favourable in normal conditions
because they reduce the velocity of ground-level wind over a considerable distance
on the leeward side and permit the deposition of snow on land stretching between the
windbreaks.

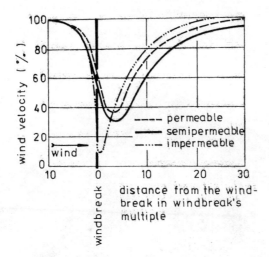

Fig. 96.   Diagram of the function of windbreaks.

Permeable windbreaks have a closed canopy which checks the wind velocity; the calm wind permeates below where there is no brush undergrowth.

Windbreaks can only be effective when established in an organic system and distributed throughout the whole protected area. Advantageous is the closed, rectangular network with the principal windbreaks perpendicular to the direction of the prevalent winds and the transverse, auxiliary windbreaks planted so as to intercept the side winds. As an example the width of the main windbreaks may be 8-11 m, 16 m in areas affected by dust storms, the height of the forest growth up to 25 m and consisting of 5-7 rows of trees. Wider windbreaks may have up to 9 or 11 rows of trees (Fig. 97).

Wind velocity is reduced by windbreaks on the windward side over a distance which is five times the height of the windbreak, on the leeward side over a distance 15-20 times the height of the windbreak. This means that the distance between the principal windbreaks in flat areas may be 20-25 times the height of the windbreak. In undulating and hilly areas the distance between windbreaks is determined by the following equation (17)

$$L = \frac{aH}{1+aI}$$ (9.13)

where L   is the distance between the principal windbreaks (m)
      H   is the height of the windbreak (m)
      I   is the slope gradient (%), expressed as a decimal number
      a   is the value of 15-25.

Trees planted in windbreaks should be fast growing pioneer tree species, with a dense canopy, adaptable to local conditions and resistant to pests and diseases. Recommended deciduous trees include: the oak, hornbeam, linden, maple, elm, birch, ash and poplar; in moist areas: aspen, krummholz and larch. The choice of trees must comply with the climatic conditions and vegetation of the area.

Fig. 97. Mechanized planting of windbreaks, South Moravia,
Czechoslovakia (photo by courtesy of the Research
Institute for Land Reclamation, Prague).

Windbreaks are extremely effective in arid and semi-arid regions where they should
be incorporated in irrigation systems. They are used for efficient erosion control
and for shading irrigation canals which they protect from silting and from the
evaporation of irrigation water. Their usefulness has been proven in many parts of
the USSR. In Japan windbreaks are used to protect the mainland from heavy winds
and fog, coming in from the sea.

9.2.5.4.2 <u>Infiltration and retention forest belts</u>. Infiltration forest belts are
planted across the slope to intercept the spring thaw and to retain it by infiltration. Soil protected by such belts does not freeze. The effectiveness of the belt
may be augmented by retention ditches, ridges and benches which are useful especially
in newly established infiltration belts.

The belts consist of a high three-storey growth with a dense brush undergrowth.
The forest belt should be dense, impermeable and the soil covered with a layer of
soft, mellow permeable litter which would quickly infiltrate water. For calculating
the width of the infiltration belt the same formula is applied as that used for the
calculation of protective strips, i.e.

$$D = \frac{oi}{w - i} L \qquad\qquad [\,m\,] \qquad\qquad (9.14)$$

where L is the length of the protected field (m) which must not exceed the
          critical length of the slope
      i is the intensity of rainfall or the average intensity of the critical
          snow thaw (m s^{-1})
      o is the runoff coefficient
      w is the infiltration rate of the soil in the infiltration belt (m s^{-1}).

The infiltration rate of the soil varies in time and space, i.e., depends on local
conditions. A favourable infiltration rate is conditional upon the presence of a
layer of humus cover.

The width of the infiltration belts is determined in dependence on the climate,
length of slope, slope gradient, and values of surface runoff. It will be 20-60 m,
at a distance of 100-600 m. The distance between infiltration belts and their
siting has been studied by M. Holý[11] who arrived at the conclusion that for
maximum effect infiltration belts should be planted in the middle of the slope
(Fig. 98) and in that part of the slope where there is a sudden change in gradient
(Fig. 99).

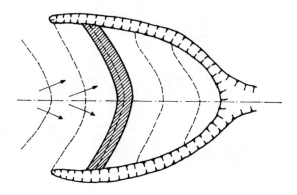

Fig. 98. Infiltration belt planted in mid-slope.

9.2.5.4.3 Shelter belts. Shelter belts shade the slopes of gullies or deep river
valleys thereby protecting and allowing the regeneration of the natural vegetative
cover. The shelter belt will reduce the temperature in the shaded area and will
create favourable moisture conditions for the growth of pioneer species. Shading
increases soil moisture, reduces the differences in temperature between the sunny
bank and the bank which is not exposed to the sun and increases the relative air
humidity at ground level.

Shelter belts consist normally of 8-10 strips of trees spaced 1.5 m apart with the
trees in the strips spaced 0.6-0.7 m apart.

The equation for calculating shading may be written

$$S = H \cot g\ \alpha \qquad\qquad [m] \qquad\qquad (9.15)$$

where S is the shaded belt (m)
      H is the height of the shelter belt (m)
      α is the position of the sun above the horizon.

The most effective shelter belts are those which are planted along narrow deep
gullies (Fig. 100). For shading wider gullies another shelter should be planted
in the lower part of the bank (Fig. 101).

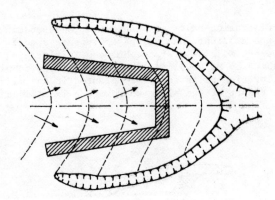

Fig. 99.   Interception strip planted at break in slope
           gradient.

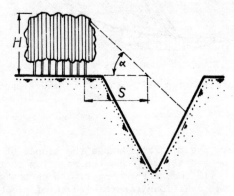

Fig. 100.   Shelter belt planted along narrow deep gully.

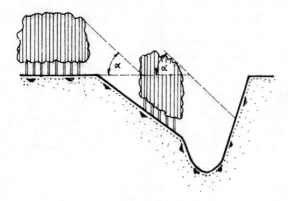

Fig. 101.   Shelter belts planted along wide gully.

Pioneer species are recommended for planting in shelter belts, such as the birch,
the aspen, the Scots pine, the oak, and quickly growing shrubs, such as the dog-rose
and hazel trees.

9.2.5.5 Afforestation

The forest is considered to be a reliable protective erosion control measure.
However, it has to be correctly established and properly managed. Only a forest
with a dense closed vertical canopy and a thick thatch of undergrowth on soil rich
with humus and with a thick cover of litter is able to fulfil its erosion control
function. Mixed forests are best for such purposes with a multi-storey growth and
adequate tree density.

The ratio of the actual state of the growth to the optimal state of the growth in
the given conditions should in mountain areas be at least 0.7-0.8 and forest
cultures with shrubs should cover at least 50% of the total area.

Erosion control protective forests should be established on hilltops and at such
altitudes where surface runoff from rain and snowfall could threaten the slopes
below. The planting of forests along the watershed and in the upper parts of
slopes is also very beneficial from the water management point of view because it
helps supply water to the fields on the slopes and in the downstream areas of river
valleys. Afforestation is recommended on the watershed and on denuded slopes with
a gradient of over 36% which can no longer be used for farming.

The beneficial erosion control effects of forest growths quickly decline with bad
management, namely excessive exploitation and the resulting disruption of the forest
soil surface and of forest paths for timber haulage. In some areas, for example,
in important water conservancy areas (catchment areas of water reservoirs), the
importance of the forest for water management is the major consideration while its
timber production function takes second place. So far, however, the management of
such forests has not yet been solved. It appears that the main forest management
technique in these forests will be clear cutting, namely in spruce monocultures with

restrictive measures, such as:

- restricting the length of clear cuttings and their immediate reforestation
- securing the network of paths and communications, i.e., the reinforcement
  of the surface of forest paths which should be built on the contours, i.e.,
  across the slopes
- building chutes for the transport of timber perpendicular to the contour
  so as to secure minimum contact between the timber and the soil surface
- limiting the unfavourable effects of mechanization in forest management
  (the replacement of caterpillar tractors by cableways, manual exploitation
  in exposed areas, etc.).

The problem of the management of multi-purpose forest growths, namely forests with
an erosion control function, is being revalued and new, efficient methods are being
sought to ensure good management and optimal economic and social utilization.

9.3 TECHNICAL CONTROL OF SHEET RUNOFF

Technical soil conservation practices reduce the intensity of erosion processes by
affecting two fundamental, morphological factors, namely slope gradient and slope
length and by creating conditions for turning surface runoff into ground water
runoff.

The classification system of erosion control practices has developed with the
gradual introduction of new types of erosion control. The line between agrotechnological, biological and mechanical measures is not sharply drawn. Technical
soil conservation practices are usually defined as measures which require besides
agrotechnological and biological, also technical measures, such as large-scale
technical improvements to the soil surface.

Technical soil conservation practices may be classified into 6 basic systems with
30 types and sub-types[23]:
   1st system - infiltration strips with grass and bush cover
      subtype : continuous strips with broad based ditches
   2nd system - cultivated broad base terraces contour and parallel types
               with installations for limiting the longitudinal movement of
               water in some cases
   3rd system - interception ditches with open ditches, lined channels and
               grated channels
      subtype : diversion and absorption ditches
   4th system - erosion control ridges - passable and impassable ridges
      subtype : diversion and infiltration ridges
   5th system - bench terraces - earth terraces, masonry terraces and terraces
               built of sections
   6th system - drainage structures.

Added to these measures should be technical projects within the catchment, such as
river training, the lay-out and construction of the communication system, irrigation
systems, etc.

9.3.1 Infiltration Strips

The function of infiltration strips planted to crops and suggestions for design are
given in section 9.2.5.2. The same method may be used for planting bush strips
which are used both as erosion control factors and as elements improving the landscape.

The effectiveness of such strips may be increased by broad-base terraces. The
strips are usually established on slopes with a slope gradient of up to 20%. The
cross profile of the broad-base terrace has a maximum gradient of 1:5 and can be
traversed with farm machines and equipment. The longitudinal slope = 0, to allow
all water flowing down from the slope to infiltrate into the grass or bush strip.
A diagram of the infiltration grass strips with broad based terraces is shown in
Fig. 102.

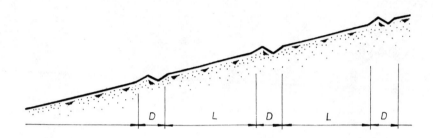

Fig. 102. Diagram of broad-based terraces.

At bigger surface runoffs on slopes, grass or bush strips with ditches are more
favourable. The ditches have a minimum bed width of 30 cm, the slope of the banks
provided the soil is cohesive is 1:1 and the longitudinal slope = 0. Above the
ditch is a 2 m wide grass strip (lucerne or clover) which protects the ditch from
being damaged by tillage operations and from silting. Suitable bushes may be
planted on the grass strip (Figs. 103a and b). This technique is extremely effective
for erosion control. On the other hand the ditches are deep and cannot be farmed
over which prevents the full use of mechanization. The system is also highly
demanding on maintenance.

9.3.2 *Cultivated Broad-base Terraces*

Soil conservation by broad-base terracing, i.e., by planting vegetation in the
broad-base ditch, consists of the interception of surface runoff (Fig. 104). Water
is absorbed into the terrace with a longitudinal slope = 0. Broad-base terraces
with a longitudinal slope divert the runoff into the outlet of the erosion susceptible area. The diversion bench is used on slopes with a gradient of up to 20%.

It applies for benches which absorb the total amount of water running off from the
slope that the capacity of the interception area must be equal to the total inflow
of water in time t at a unit width of the broad-base terrace, thus

$$hb_s = oitL \tag{9.16}$$

where b_s is the average width of the terrace (m)
      h is the depth of the terrace (m)

o is the runoff coefficient
i is the intensity of rainfall (m s^{-1})
t is the duration of rainfall (s)
L is the spacing of terraces (critical slope length - see section
   9.1) (m).

The horizontal distance between two terraces may be derived from equation (9.16)
and written

$$L = \frac{hb\underline{s}}{oit} \qquad [m] \qquad\qquad (9.17)$$

The vertical interval between two terraces may be derived from the same equation
and written

$$H = LI \qquad\qquad (9.18)$$

where I is slope gradient.

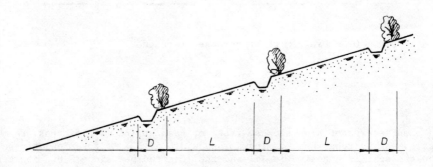

Fig. 103a. Grass strips with channels.

The smallest recommended depth of terraces is 0.5 m, bank slopes 1:5 (or better
1:10) to allow machinery to traverse and the minimum area of the cross profile
0.8 m^2. Standard rain is usually selected with periodicity p=0.1, duration t=15
mins and runoff coefficient o is obtained from the equation given in section 3.1.4.

Diversion benches which are established in less permeable or impermeable soils have
a longitudinal slope of 1-5‰. They divert runoff into the outlet.

Cultivable broad-base terraces are established along the contour or are parallel and
installations may be built to limit the longitudinal movement of water. The choice
of terrace is given by the configuration of the terrain. Contour terraces are
favourable for areas with a complex configuration which does not allow the establishment of parallel terraces which are more widely used to preserve the regular shape
of the plot. The establishment of parallel broad-base terraces in Czechoslovakia
in the South Moravian region is shown in Fig. 105[6].

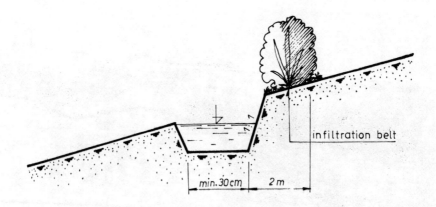

Fig. 103b. Channel with an infiltration belt.

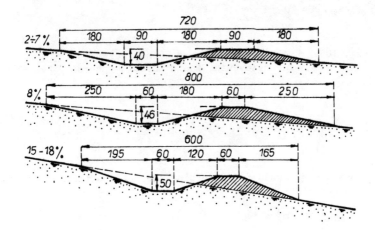

Fig. 104. Cultivated broad-base terraces.

9.3.3 Interception Ditches

Interception ditches are designed for areas with a slope of up to 20% and extremely susceptible to erosion. They intercept and divert surface runoff and enhance its infiltration into the soil. These ditches are classified into diversion and absorption ditches.

Fig. 105. Establishment of broad-base terraces in
Czechoslovakia, South Moravia (photo by E. Dýrová).

9.3.3.1 Diversion ditches

Diversion ditches are designed for diverting the outflow from a catchment. Flow
rate capacity may be written

$$Q = Pq = Fv \ (m^3 \ s^{-1}) \tag{9.19}$$

where P is the area of the catchment (ha)
      q is the specific surface runoff from the catchment $(m^3 \ s^{-1} \ ha^{-1})$
      F is the area of the ditch bed (m^2)
      v is the mean flow rate $(m \ s^{-1})$

Runoff from torrential rainfall is the determinant for calculating the specific
inflow from catchment q. Usually torrential rain is chosen for the calculation
with a periodicity of p = 0.1. The runoff coefficient may be calculated using any
of the methods given in section 3.1.4. For small catchments with an area not
exceeding 100 ha the retardation of the inflow and the variability of rainfall
intensity are usually not considered.

A steady flow is assumed in the ditch. The calculated mean profile velocity should
not exceed maximum permissible friction velocity. Its values, empirically derived
by many authors, are given in the literature(14).

For purposes of design the ditch will be divided into sections and to each section will be assigned the respective catchment P_1, P_2 .... and the corresponding flow profile F_1, F_2 (Fig. 106).

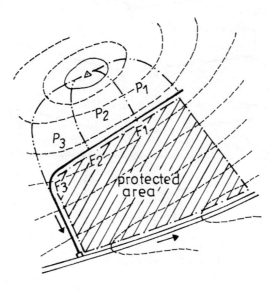

Fig. 106.   Area protected by a ditch.

For the erosion control of larger areas a series of diversion ditches should be designed running into a channel and from there into the recipient (Fig. 107).

Ditches are usually trapezoidal and the interception area is extended by a ridge on the lower end of the slope of the dug-up earth (Fig. 108).

The banks of the ditches and ridges are grassed.

V. Sedlák[23] designed for Southern Moravia in Czechoslovakia ditches which are built of prefabricated concrete segments and covered with grates (Fig. 109).

The advantage of this system, which is very useful for groves and vineyards, is that farm machines can travel in all directions. The disadvantage is that a perfectly functioning water diversion system has to be built requiring higher capital costs. The return of the means expended was 7 years with groves, and 10 years with vineyards. This showed the benefits of using the grated ditch system in special conditions.

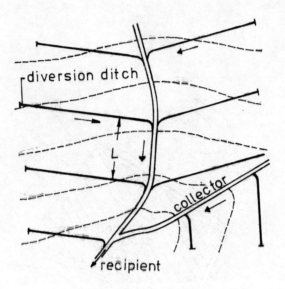

Fig. 107. A system of diversion ditches.

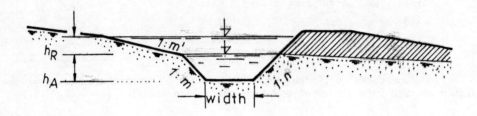

Fig. 108. Detail of a diversion ditch.

9.3.3.2 Absorption ditches

Absorption ditches intercept the total amount of water flowing from higher parts of the area and will retain it until it is absorbed into the soil. They are designed to take the whole volume of the inflow.

Fig. 109. Ditch of prefabricated concrete segments covered with grates (photo by V. Sedlák).

9.3.4 *Erosion Control Ridges*

The erosion control effectiveness of the system consists of intercepting surface runoff by a series of low ridges (Fig. 110). Water intercepted by diversion ridges designed with a longitudinal slope of up to 10% is carried out of the catchment while water intercepted by absorption ridges with a zero longitudinal slope is absorbed by the soil. The ridges are either traversable with a slope gradient of a minimum of 1:5 or intraversable with a slope gradient of 1:1.5.

Fig. 110. Erosion control ridge (photo by courtesy of the Institute of Irrigation and Drainage, Technical University, Prague).

The cross section of the ridges and that of the ditches must be in agreement with the area of interception.

Diversion ridges are designed for heavier soils with low infiltration capacity. Their length should allow the maximum quantity of water to be diverted during rainfall. The length of the ridge conforms to this demand

$$D = vt \quad (m) \tag{9.20}$$

where v is the velocity of outflow along the ridge $(m\ s^{-1})$

$$v = c\ \sqrt{RI'}$$

where c is the coefficient of velocity $(m^{1/2}\ s^{-1})$
$\quad$ R is the hydraulic radius of the flow profile behind the ridge (m)
$\quad$ I' is the longitudinal slope of the ridge
$\quad$ t is the duration of rainfall (s).

Usually the length of the ridge is 300–450 m, occasionally more. On impermeable soils a ditch is sometimes cut along its upslope heel to improve outflow.

Absorption ridges should not overflow. This condition is the criterion for investigating the dimension of ridges and their spacing (Fig. 111). It applies that

$$\frac{h}{2} \left(\frac{h}{I} + hn \right) = oitL \tag{9.21}$$

where h is the height of the ridge (m)
$\quad$ n is the bank slope (usually $n = 1$ to 4)
$\quad$ I is the slope gradient of the protected slope (%)
$\quad$ i is the average intensity of rainfall $(m\ s^{-1})$
$\quad$ t is the duration of rainfall (s)
$\quad$ o is the outflow coefficient
$\quad$ L is the spacing of ridges, not exceeding the critical slope length (m).

The spacing of ridges may be expressed as

$$L = \frac{h^2}{2oit} \left(\frac{1}{I} + n \right) \quad [m] \tag{9.22}$$

the height difference between the ridges may be written

$$H = LI = \frac{h^2}{2oit} \left(1 + nI \right) \quad [m] \tag{9.23}$$

9.3.5 *Bench Terraces*

Bench terraces are established on deep soils and steep slopes, with a slope gradient of more than 15%, which intercept, retain and possibly divert surface runoff. The shape of bench terraces and their height depends on the slope of the area, the depth of the soil profile, soil levelling, access to mechanization, method of cultivation, etc.

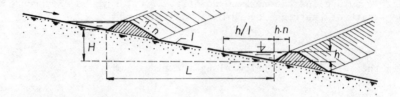

Fig. 111. Dimension of ridges and their spacing.

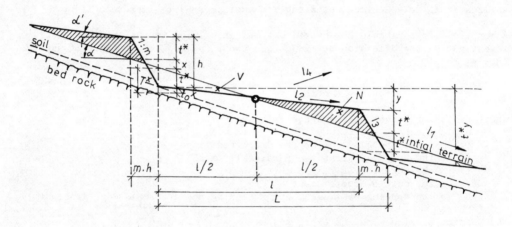

Fig. 112. Design of terrace benches after K. Kasprzak.

Terrace benches may be designed after K. Kasprzak[13] (Fig. 112). In this figure

T^* is the depth of the soil profile to bed rock (m)
t_0^* is the depth of the soil profile usable for vegetation 0.6–0.8 m
t^* is the permissible depth of excavation for terrace building (m)
V is the excavation (m^3)
N is the fill – it applies that V = N
ℓ is the width of the terrace platform (is bisected by a point of intersection with the initial terrain) (m)
h is the height of the terrace bench (is bisected by the point of intersection with the initial terrain) (m)

L is the width of the terrace (must not exceed permissible slope length (m))
I_1 is the average slope of the original terrain (%)
I_2 is the designed slope of the terrace (%)
I_3 is the slope of the terrace bench dependent on soil and reinforcement
   (is expressed by the ratio 1:m)
I_4 is the longitudinal slope of the terrace (dependent on the configuration
   of the terrain, soil permeability, precipitation conditions) (%)

From Fig. 112 may be derived equations for ascertaining the height of the terrace
platform

$$h = \frac{2(T^* - t_0^*)}{\ell - I_1 m} \qquad [m] \qquad\qquad (9.24)$$

the width of the terrace platform

$$\ell = \frac{2(T^* - t_0^*)}{I_1 - I_2} \qquad [m] \qquad\qquad (9.25)$$

the designed slope of the terrace platform

$$I_2 = I_1 - \frac{2(T^* - t_0^*)}{\ell} \qquad [\%] \qquad\qquad (9.26)$$

and the total width of the terrace

$$L = \frac{2(T^* - t_0^*)}{I_1 - I_2} + hm \qquad [m] \qquad\qquad (9.27)$$

Bench terraces are built to intercept surface runoff from the whole terrace area
with width L or to secure a non-erosive outflow of runoff into the outlet at slope
gradient $I_4 > 0$ (Fig. 113).

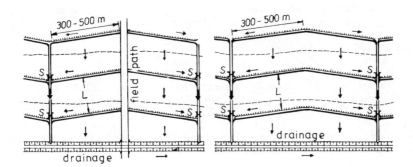

Fig. 113. Diagram of the function of bench terraces.

The interception area of the terrace may be arranged in three ways:
   - by a positive slope of the terrace platform $I_2 > 0$; water from the terrace
     bench will be absorbed by the soil or be non-erosively carried away;
     infiltration may be enhanced by furrows cut along the whole terrace platform
     and water may be diverted by cutting a ditch at the lower end of the
     platform (Fig. 114a)

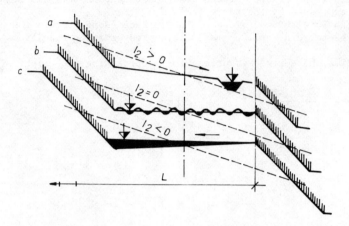

Fig. 114a,b,c. Types of bench terraces.

  - by the establishment of a level terrace platform $I_2 = 0$, intensive
    infiltration may be attained by furrows, as above (Fig. 114b)
  - by a negative slope of terrace platform $I_2 = 0$; the interception area must
    retain the entire outflow of water from the area with width L (Fig. 114c).

In less permeable soils the terraces must be built with a mild longitudinal slope
I_4 in order to divert water from the whole terrace platform.

The establishment of earth terraces is relatively costly. Earth terraces will,
however, be formed spontaneously along the contours of contour cultivated fields
with boundaries, provided suitable agrotechnology is applied (Fig. 115).

Stone wall terraces allow water to seep through the stone wall and prevent it from
accumulating behind the terrace bench (Figs. 116 and 117).

In hilly terrain planted with permanent valuable cultures, such as groves and
vineyards, sections are used (Figs. 118a,b and 119) for building terraces. The
bench terrace then becomes a highly demanding structure. The platform of the
terrace sections is square with a sloping front edge to the bench or with sloping
front and back edges. The slopes of terrace sections have a slope gradient of 1:1
and are covered with vegetation. Productivity on terrace sections comes close to
the productivity attained on mild natural slopes and on level land. Investment
costs are increased by the necessity to build a perfectly functioning drainage
network. As an erosion control element the terrace sections do not blend with the
surrounding countryside too well. For the conditions of South Moravia there is a
total return on capital after 11 years.

The use of terraces in Czechoslovakia is shown in Figs. 120, 121 and 122 and in China
in Fig. 123. The use of graders for building terraces in the US is shown in
Fig. 124.

Fig. 115. Spontaneously formed terraces (photo by J. Říha).

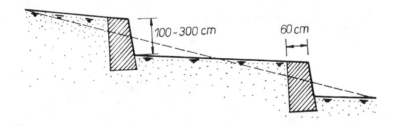

Fig. 116. Stone wall terraces.

Fig.   117.   Stone wall terraces in vineyards (photo by
             M. Holý).

CROSS   SECTION

Fig. 118a.   Bench terrace in vineyards.

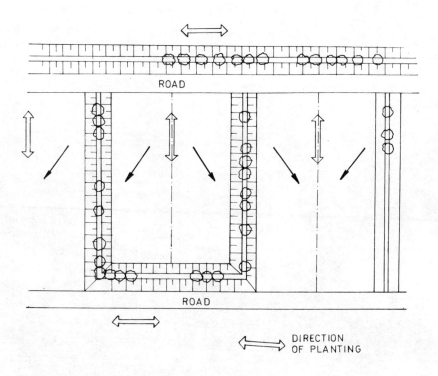

ROAD

ROAD

DIRECTION
OF PLANTING

Fig. 118b. Bench terrace in vineyards.

9.3.6 Drainage Structures

Erosion control requires that all surplus water that would cause erosion be non-
erosively diverted from the area. For this purpose it is necessary to build drainage
structures.

9.4 TECHNICAL CONTROL OF CONCENTRATED SURFACE RUNOFF

Concentrated runoff is evident by gully erosion which culminates in fluvial erosion.
Unless effective measures are taken fluvial erosion will devastate the beds of water
flows (Fig. 125), especially of streams and the adjoining area, communications,
buildings (Fig. 126), etc. Gully erosion should be prevented by surface runoff
control in the catchment using a suitable combination of agrotechnical, biological
and technical measures. The consequences of fluvial erosion may be avoided by
technical measures supplemented by the planting of suitable vegetation.

Technical measures for the control of concentrated runoff include:
  - erosion control reservoirs
  - the reclamation of gullies and ravines
  - torrent control.

Fig.119. Bench terrace (photo by V. Sedlák).

9.4.1 Erosion Control Reservoirs

Erosion control reservoirs, mostly ponds, fulfil the following four fundamental
functions:
- they intercept any sudden impact of surface runoff thereby preventing gully
  erosion in the area below the reservoir
- they intercept sediments
- they raise and stabilize the erosion base of the respective catchment area
- they improve the moisture regime of the soil and atmosphere thereby
  increasing the erosion resistance of soils susceptible to erosion.

Erosion control reservoirs may be temporary or permanent. Temporary reservoirs
are not restored after silting, the silted area is cultivated into fields, meadows
or forests. Permanent reservoirs function not only as erosion control units but
also enhance the further use of the area which depends on the time course of

silting. At a certain sediment silt level the sediments must be removed to allow
the reservoir to fulfil its function.

Fig. 120. Terraces in Czechoslovakia, South Moravia
(photo by V. Sedlák).

The erosion control function of reservoirs depends on their location, technical
layout, operation and maintenance. To be able to intercept the dangerous outflow
from the catchment the reservoirs must be correctly established (Fig. 127).
Reservoirs at the highest points in the catchment where concentrated surface runoff
originates and initiates dangerous intensive gully erosion, intercept torrential
rain and spring snowmelt and subsequently carry the runoff into reservoirs at lower
points in the catchment which are mainly used for storage and other economic
purposes. A correctly established system of reservoirs helps balance the outflow
from the catchment thereby protecting it from the effects of erosion. These
reservoirs will also improve the water supply for industry and agriculture in the
respective area.

Fig. 121.   Terraces in Czechoslovakia, South Moravia
(photo by V. Sedlák).

Fig. 122. Aerial view on terraces (photo by E. Dýrová).

Very important in stream perimeters are gravel collectors (Fig. 128) which serve
to intercept bed-load sediments carried by the stream.

9.4.2 Control of Gullies and Ravines

Gullies and ravines result from the intensive scouring action of concentrated surface
runoff. Precipitation water concentrates in the heads of gullies and ravines and
gradually cuts into the slope thereby enhancing the growth of the gully or ravine.

Gully and ravine control may be carried out:
   - by gradual treatment
   - by checking.

Fig. 123. Terraces in China (photo by courtesy of Institute
of Irrigation and Drainage, Technical University,
Prague).

Gradual treatment is the method used to control deep gullies and ravines mostly on
land not used for farming. It consists of the protection of the gully head bank,
bed treatment and slope stabilization. It is carried out in stages, the first
stage being the prevention of the lengthening and deepening of the gully or ravine,
the second stage being the stabilization of slopes, usually by vegetation.

The gully and ravine heads are protected by interception ditches with an advance
check built either above the gully head (Fig. 129) or checks built in series
(Fig. 130). The intercepted surface runoff which is not absorbed by the soil is
diverted into the gully (ravine) by an inlet culvert secured against scouring. A
series of erosion control measures consisting of agrotechnical, biological and
technical measures should be implemented to reduce gully head overflow (sections
9.2, 9.3).

Fig. 124. Building of terraces in USA (photo by courtesy
of the Soil Conservation Service, USA).

Treatment of the gully or ravine bottom should ensure the non-erosive diversion of
the water. The technique of the treatment is determined by the quantity of water and
its flow in time, the longitudinal slope of the gully bottom, the shape of the cross
profile, the state of the gully bottom, the soil properties and the purpose of the
treatment.

Small shallow gullies not affected by continuous flow may be stabilized by continuous
vegetation and suitable grasses, bushes and trees. Erosion control of deep gullies
and ravines continuously affected should be carried out by technical means, namely by
terraces and check dams (Fig. 131).

Fig. 125. Devastated bed of the Váh river tributary in
Czechoslovakia (photo by J. Říha).

Appropriate checks will control gully overflow, intercept silts and will gradually
transform the longitudinal gradient of the gully bottom from a steep slope to a
steady-state slope of 0.5-1%, i.e., one that will correspond to the outflow of clean
water. This is achieved by bottom treatment which will in its first stage form a
compensated slope gradient of 5-10%. In the first stage a primary series of checks
will be put in, later secondary checks will be built midway between the primary
checks.

The stabilization of gully and ravine slopes is to prevent their further scouring
and undercutting.

The basic measure to be taken is the stabilization of the foot of the slope. Variou
techniques are used depending on the erosion susceptibility of the slope. Less
exposed slopes are stabilized by vegetated structures, plank fences, vegetated rubbl
checks, etc. Embankments should be built on highly exposed sites.

Fig. 126.   Devastating effect of stream (photo by J. Říha).

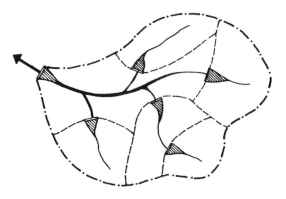

Fig. 127.   System of reservoirs in catchment.

Fig. 128. Gravel collector in North Bohemia (photo by
           J. Říha).

Single treatment is only used for deep gullies and ravines and sunken roads dividing
fields into small runs. Suitable mechanizations should be used for filling these
gullies, ravines and sunken roads. It is necessary to prevent concentrated runoff
in the area.

9.4.3 Torrent Control

Concentrated surface runoff causes erosion of various intensity affecting the banks
and beds of torrents and other water flows and causing the formation and transportation of sediments. This process may be prevented or alleviated by stabilizing the
bed, lining the banks and by other techniques.

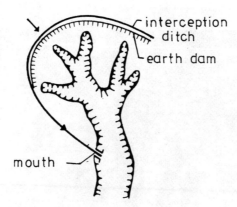

Fig. 129.   Protection of a ravine.

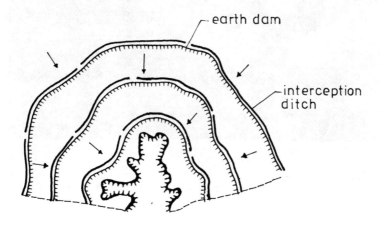

Fig. 130   Protection of a ravine.

The objectives of torrent control include:
- the treatment of the stream perimeter such as would affect the concentration
  of runoff in the stream and would prevent the excessive transportation of
  sediments from the perimeter to the stream bed
- the treatment of the stream bed slope to reduce the scouring action of the
  water flow

- the treatment of the flow profile of the stream to secure non-erosive flow.

Fig. 131.   Check dams in a ravine (photo by courtesy of the
            Institute of Irrigation and Drainage, Technical
            University, Prague).

9.4.3.1 Treatment of stream perimeter

Unfavourable flow conditions in the stream arise due to the sudden concentration of
water flowing into the stream from the perimeter during torrential rain or sudden
snow thaw. Maximum runoffs may be expected from perimeters with steep slopes, low
soil infiltration capacity and a low retention capacity of the soil surface which
usually has an inadequate vegetative cover. It follows that erosion control
measures should be aimed at intercepting surface runoff, changing most of it into
groundwater runoff and stabilizing the soil surface.

In the cultivated part of the perimeter erosion control agrotechnology should con-
sistently be implemented, contour cropping introduced, erosion control crop
management generally applied and various types of infiltration strips and terraces
established, provided this measure is economic in the given area. All technical
projects in the perimeter should be implemented strictly observing the gradient of
the area - this aspect should especially be borne in mind in laying out road systems,
irrigation and drainage systems, etc.

Steep slopes on which farming is uneconomical even with erosion control should be
protected with a permanent grass or forest cover.

The treatment of the stream perimeter is extremely important because suitable
measures organically integrated in the economic and technical arrangement of the
area can significantly contribute to effective and economic torrent control.

9.4.3.2 Stream bed treatment

In order to stabilize the stream bed the longitudinal slope of the stream bed will
be levelled into a compensation slope; in streams with an irregular bed slope this is
done by a series of benches, checks and stabilization strips. Consolidation checks
are built to grade the stream bed and to reduce the longitudinal slope by intercepting and stabilizing sediments; retention checks are built to intercept sediments and
to prevent their further transportation.

Consolidation checks consolidate sediments along the so-called sediment line drawn
between the crown of the lower check with the toe of the nearest upstream check.
A check of height H (Fig. 132a) will after a certain period of time form sediment
line with gradient I´ and range D.

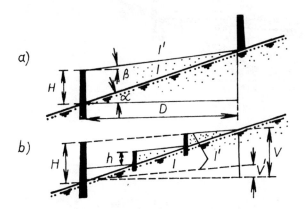

Fig. 132a and b.   Schemes of consolidation checks.

The equation may be written

$$D = \frac{H}{tg\alpha - tg\beta} = \frac{H}{I - I´} \qquad [m] \qquad\qquad (9.28)$$

where I is the slope of the initial stream bed
      I′ is the slope of the sediment line and thereby of the new stream bed.

The height of the greater number of lower checks built to replace higher checks
is shown in Fig. 132b and may be written

$$h = d(I - I') \qquad [m] \qquad (9.29)$$

where $d = \dfrac{D}{n}$ and n is the number of checks.

In order to grade the stream bed by consolidation checks primary checks are first
built which will grade the longitudinal slope of the stream bed into a compensated
slope and to these primary checks will then be added secondary checks which will
level the longitudinal profile of the stream into a steady state slope (Fig. 133).

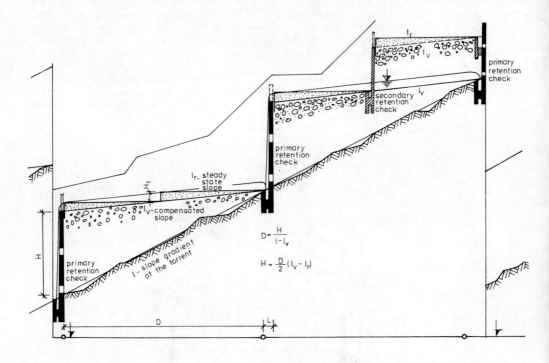

Fig. 133. Consolidation of a torrent.

Retention checks are built to retain course and boulder sediments while fine sediment
which would soon silt the retention area should be washed out. These fine grained
sediments are washed out by water flowing over the spillway, at lower water levels
by water flowing through apertures at different heights in the dam body.

Retention checks should have a large retention storage and are built in areas which will have such a retention storage for a relatively small width of dam wall which may rise to a considerable height. Masonry retention checks and dams are often 10 m in height and more. The retention storage is designed to comply with the regime of silting and the type of sediments.

The statistical calculation of checks and dams involves securing safety against silting and sliding, observing permissible stress in the masonry and in the foundation joint as well as permissible load on foundation ground.

Examples of retention checks are given in Figs. 134 and 135a and b.

Cross structures lower than checks are built which grade the longitudinal slope of the stream to a compensation slope, possibly to a steady-state slope. They are made of various materials. Examples are shown in Figs. 136, 137 and 138.

In some cases boulder chutes are built instead of steps. These chutes are less costly, hydraulically more effective and blend with the surrounding countryside. Examples of such chutes are shown in Figs, 139, 140 and 141a and b.

Benches may be built to make small alterations to the stream bed. They are built of stone masonry, concrete, prefabricated concrete segments and wood. Examples are shown in Figs. 142 and 143.

The construction of benches is advantageous especially in smaller projects. They blend with the surrounding countryside better than do higher structures. On the other hand they are hydraulically ineffective for high flow rates because the excess energy of the water is not damped above the stilling basin.

Stream bed checks are built with crowns at stream bed level to prevent bed scouring in streams with unlined beds. In streams with lined beds they prevent the outflow of water beyond the lining. They are mostly built of stone, wood and concrete and lined with stones in the flow area (Fig. 144).

9.4.3.3 Treatment of flow profile

The stream flow profile should secure the nonerosive flow of water.

On the upper reach of streams, where the bed is usually cut deep into the bed rock, treatment of the cross profile shape is extremely limited. Treatment is restricted to the gradient. On the middle and lower reaches of streams it is necessary to alter both the direction of the flow and the cross section of the stream profile.

Stream bed treatment depends on the size, frequency and character of the flow. With adequate bed capacity it suffices to improve the flow conditions by correcting the sharp bends and turns in the stream bed and by removing obstructions, such as boulders, stone debris which has fallen from the surrounding slopes, upturned trees, etc. Damaged stream banks should be repaired and the vegetative cover completed.

The routing of the stream bed should secure a continuous flow. The cross profile of the stream bed is usually designed so as to prevent torrential waters from over-flowing the stream bed and spreading into the surrounding area. The extent of the possible flooding is determined by the configuration of the terrain, land use, the vicinity of structures, etc. The most frequently designed cross profile is trapezoidal, with a corresponding lined bank slope. A narrow profile is justified in those parts of the stream where the transportation of sediments must be secured requiring greater tangential stress of flowing water. In wide stream beds suscep-tible to gravel silting a new stream bed may be obtained by spurs built either transversely or along the designed route of the stream bed.

Fig. 134. Retention checks in the Alps (photo by J. Říha).

Fig. 135a.   Retention check in North Bohemia (photo by J.
              Pretl).

Vegetation and vegetated structures are widely used for damming streams. It should,
however, be borne in mind that growing vegetation may considerably change the
hydraulic properties of the bed. Seeding or sodding will change the roughness of
the flow area. Vegetated structures and high bushes and trees will not only change
the roughness of the bed but will also narrow the flow profile. This should be
considered in hydraulic calculations of the stream bed.

The choice of vegetation used for lining stream beds (Fig. 145a and b) depends on the
permissible water velocity, on water depth and on the permissible cure of continuous
flooding.

Highly exposed stream beds should be lined with paving (Fig. 146) sometimes in
combination with wood, logs or stone walls (Fig. 147).

Stream bed treatment must always be conceived as part of outflow control in the
stream catchment and as part of complex erosion control.

Fig. 135b. Retention check in North Bohemia (photo by
            J. Pretl).

9.4.3.4 Design of stream bed

The main task in the design of stream beds is to determine the cross profile and
longitudinal slope. In order to secure stream bed stability, the mean profile
velocity of water corresponding to the standard flow rate must be lower than the
maximum permissible velocity of water measured on the stream bed. As it is difficult
to measure and express water velocity on the stream bed attempts have been made to
convert it to mean friction velocity. For deriving it various authors have applied
a number of simplifying assumptions based on local conditions. Before applying the
equation the conditions for which it had been derived should be ascertained and
compared with conditions in the locality studied.

The proven equations derived by N. V. Goncharov and E. Meyer-Peter may be applied
(cit.19) for calculating the mean friction velocities of water in noncohesive soils.

N. V. Goncharov gives the equation for grain sizes with a mean diameter of 0.1-1.5
mm. The equation may be written

$$v_v = 3.9h^{0.2} \left(\frac{d_s}{d_{90}}\right)^{0.2} \left(d_s + 0.0014\right)^{0.3} \tag{9.30}$$

Fig. 136. Cross structure in North Bohemia (photo by
           M. Holý).

For grain sizes with a mean diameter of 1.5-20 mm the equation is written

$$v_v = \log \frac{8.8h}{d_{95}} \sqrt{2g \frac{\rho_m - \rho}{1.75\rho}} \; d_s \qquad\qquad (9.31)$$

where v_v is the mean friction velocity of water (m s^{-1})
      h is the water depth (m)
      d_s is the mean diameter of grains of material on the stream bed (m)
      d_{90}, d_{95} is the diameter of grains of material on the stream bed, corresponding
            to 10% or 5% respectively of the weight residue intercepted by screens (m)
      g is acceleration due to gravity (m s^{-2})
      ρ_m is the specific weight of sediments (t m^{-3})
      ρ is the specific weight of water (t m^{-3}).

For grains with a grain size of more than 20 mm the expression may be applied for
mean profile velocity derived from the equation for the movement of sediments
derived by E. Meyer-Peter and written

$$v_v = 5.77h^{1/6} \; d_s^{1/3} \qquad\qquad (9.32)$$

The values of scouring velocities calculated from equations (9.30, 9.31, 9.32) are
given in graphs shown in Figs. 148, 149 and 150.

Fig. 137. Cross structure at Elbe river in Giant Mountains,
Czechoslovakia (photo by J. Říha).

Having obtained the standard flow rate Q_n, the mean grain size of material on the
stream bed d_s and knowing the determined mean profile velocity it is possible to
design the cross profile of the stream bed.

Profiles of gravel beds should be more shallow and wider with a width-depth ratio
of 8 to 10. In such case there will be good agreement between real and calculated
v_v.

The longitudinal slope of the stream is usually designed in the first stage as a
compensation slope. Many formulas for its calculations may be found in the
literature. E. V. Thiéry (cit.22) founds his calculation of bed stability on the
condition that mean friction velocity can only reach the values of bed flow
velocity. On this assumption he derived the following equation:

$$I = \frac{b(\rho_o - \rho) \; f\cos\alpha}{0.076 \; \rho c^2 R} \qquad [\,\%\,] \qquad\qquad (9.33)$$

Fig. 138. Cross structures in a village (photo by J. Říha).

where I is compensation slope (%)
      b is the size of boulders (m)
      ρ_0 is the specific weight of silt (t m^{-3})
      ρ is the specific weight of water (t m^{-3})
      f is the coefficient of friction
      α is the gradient of the stream bed
      c is the coefficient of velocity after Bazin (m$^{1/2}$ s^{-1})
      R is the hydraulic radius (m).

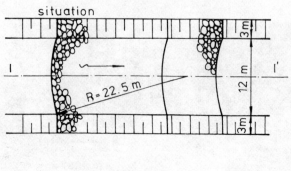

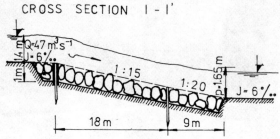

Fig. 139. Boulder chute.

C. R. Valentini[28] simplified the equation when testing Thiery's relation in the North Italian Alps and wrote it

$$I = 0.093 \frac{a}{R} \qquad [\%] \qquad (9.34)$$

where a is the dimension of the edge of a cube which by its volume corresponds to the volume of the average mean grain d_s (m)
   R is the hydraulic radius (m).

E. Meyer-Peter (cit. 22) derived a formula for calculating the compensation slope whose modified and simplified form may be written

$$I = 0.075 \ 2 \frac{d_s}{h} \qquad [\%] \qquad (9.35)$$

where d_s is the diameter of the mean grain (mm)
   h is the depth of the water (m).

The compensation slope should be calculated using more than one formula. That value should then be applied which approaches most closely the greatest number of results obtained.

The design of the cross profile and the compensation slope should be assessed to see that they apply for mean velocity $v<v_v$ and the required Q.

Fig. 140.   Concrete chute (photo by M. Holý).

The conditions of flow often change in streams with a slope of more than 3% owing to the aeration of the water. In this case the formula written by R. Ehrenberger[7] for calculating mean velocity will be applied:

$$v_v = v_1 k_p \qquad\qquad [\,m.s^{-1}\,] \qquad\qquad (9.36)$$

where $v_1 = 55\ R^{8.52}\sin\alpha^{0.40}$

      R is the hydraulic radius (m)

α is the gradient of the stream bed
k_p is the coefficient of aeration
        for α > 28°30' $k_p = 0.28R^{-0.05} \sin\alpha^{-0.74}$
        for α < 28°30' $k_p = 0.40R^{-0.05} \sin\alpha^{-0.26}$

Auxiliary values for calculating k_p and v_1 after R. Ehrenberger are given in Tables 9.3 and 9.4.

Fig. 141a and b. Boulder chute.

Fig. 142. Wood cross structure (photo by J. Pretl).

Ehrenberger's formula is derived for a wooden trough where $\gamma = 0.06$ and the respective roughness should therefore be introduced into the formula for other materials.

The equation written by Ničoporovič for steep slopes also applies for stone stream beds

$$v_v = v_1 \ k'\qquad\qquad\qquad (9.37)$$

where v_1 is $37 \ R^{0.52} \ I^{0.40}$ (m s^{-1})
      k' is $0.45 \ R^{-0.05} \ I^{-0.26}$
      R is the hydraulic radius (m)
      I is the bed gradient (m)

The equation applies for $\gamma = 0.46$, $R \leqq 0.30$, $I < 0.477$.

Fig. 143. Wood cross structures (photo by V. Zelený).

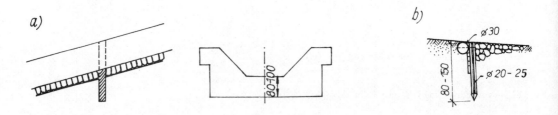

Fig. 144. Stream bed checks.

Fig. 145a. Vegetation lining of a stream (photo by
            D. Zachar).

Fig. 145b. Vegetation lining of a stream (photo by M. Holý).

9.5 CHEMICAL EROSION CONTROL

In the process of the construction of various engineering projects, such as water
projects, roads and other communications, housing estates, as well as land reclamation,
slopes are often formed which are affected by intensive erosion in the period between
project completion and stabilization by vegetative cover or by other means. Worst
affected are the banks of earth dams, motorways, mine tips etc. The surface of such
slopes may be stabilized by chemical means which will provide temporary erosion
control for the time required for the implementation of permanent measures, namely
the planting of the vegetative cover.

Chemical substances used for such purposes are varied.

An important group are polymers and copolymers inducing processes in the soil which
result in the formation of stable aggregates. Aggregation is achieved by reducing
the electrokinetic potential on the surface of soil particles and the adsorption of
polymeric molecules which is initiated by the growth of active forces.

Aggregation processes are initiated by polymers of acrylic derivates (acrylic acid,
acrylamide, methylacrylamide, polymers of sulfone polysterene, polyvinylalcohol), and
polymers dissoluble in water which are prepared by the hydrolysis of polymerized
esters (polyvinyl acetate, acrylonitrile, methacrylate, etc.).

Fig. 146. Lining with paving (photo by J. Vondrák).

These preparations are applied to the soil in different concentrations whose optimal values are tested experimentally. The process of the sorption of the preparation and aggregation has a high initial velocity and 75% of the process will take place within 24 hours.˰8).

The properties of soils with an artificial aggregation of soil particles are the same as those of soils with natural aggregation, the soils have a favourable water and air regime. The applied chemicals have no toxic effect and have no adverse effect on the nutrient regime of the soils. With regard to its high erosion resistance, artificial aggregation of soils is mainly used in road building and in water management projects.

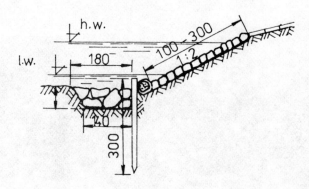

Fig. 147.   Lining with paving (photo by J. Vondrák).

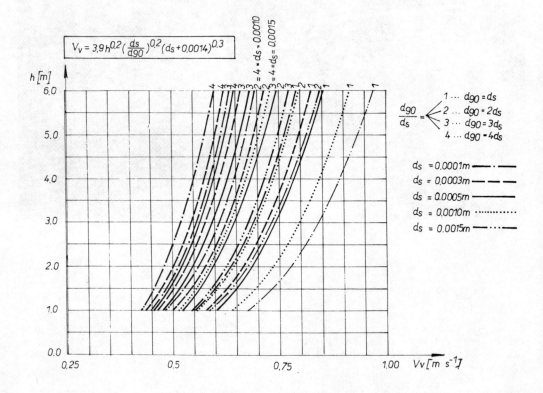

Fig. 148.   Values of scouring velocities calculated from
            equation (9.30).

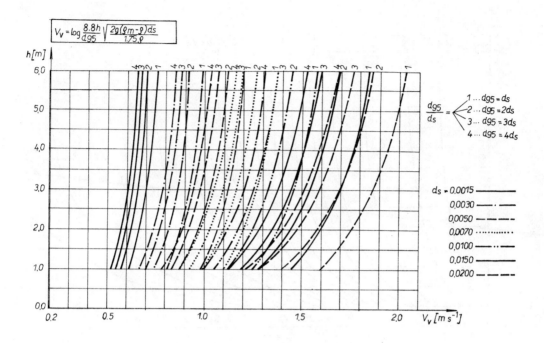

Fig. 149.   Values of scouring velocities calculated from
           equation (9.31).

There is a wide range of chemicals for the aggregation of soil which are marketed
under different trade marks. Krilium 6, Separan 2610 and HPAN are US products,
K4, K5, K6 and Nerozin are made in the USSR, Curasol AE and Curasol AH in the FRG,
VAN in the GDR, PAM and Ucedek in Belgium, etc.

The application of polymeric preparations for water and wind erosion control
considerably reduces the intensity of erosion processes[8,27 etc.).

Recently the possibilities of applying chemical mulching substances are being widely
studied. These preparations do not affect the soil heterogeneous dispersion system.
These highly adhesive and flexible substances form a protective film on the soil
surface without having any adverse effect on the soil moisture, air, heat and
nutrient regime.

These substances include crude oil derivatives, namely asphalt, saturated and
unsaturated aliphatic hydrocarbons from crude oil, natural gas and tar (polyethelene,
resins, benzooxazine, polyolefin). A separate group comprises synthetic rubber,
latexes, latex oil, emulsions and bitumens.

These substances have excellent heat stability and resistance to oxidization. They
facilitate the cementing of soil surface pores, the soil particles acquire hydro-
phobic properties and the loss of water by evaporation is reduced.

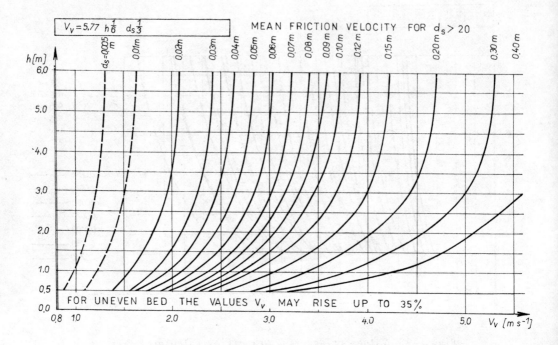

Fig. 150.   Values of scouring velocities calculated from
            equation (9.32).

The wide range of substances produced worldwide include Unisol 91g (latex oil
emulsion), made in the FRG, Humofina B 2864 (asphalt emulsion) made in Belgium,
Shell Solfix (bitumen) made in Holland, $L_x PO$ (latex emulsion) made in
Czechoslovakia, etc.

Shell Solfix is recommended for the stabilization of light soils[24]. Bitumen based
substances have successfully been used for the stabilization of sand dunes[8].

TABLE 9.3   Function of hydraulic radius R after
R. Ehrenberger.

R	$R^{0.52}$	$55R^{0.52}$	$R^{-0.05}$	$0.4R^{-0.05}$	$0.28R^{-0.05}$
0.1	0.30	16.61	1.11	0.44	0.31
0.2	0.43	23.82	1.08	0.43	0.30
0.3	0.53	29.43	1.06	0.42	0.30
0.4	0.62	34.16	1.05	0.42	0.29
0.5	0.70	38.34	1.04	0.41	0.29
0.6	0.77	42.19	1.03	0.41	0.29
0.7	0.83	45.71	1.02	0.41	0.29
0.8	0.89	49.01	1.01	0.40	0.28
0.9	0.95	52.09	1.01	0.40	0.28
1.0	1.00	55.00	1.00	0.40	0.28
1.1	1.05	57.81	1.00	0.40	0.28
1.2	1.10	60.50	0.99	0.40	0.28
1.3	1.15	63.03	0.99	0.39	0.28
1.4	1.19	65.51	0.98	0.39	0.28
1.5	1.24	67.93	0.98	0.39	0.27
1.6	1.28	70.24	0.98	0.39	0.27
1.7	1.32	72.49	0.97	0.39	0.27
1.8	1.36	74.64	0.97	0.39	0.27
1.9	1.40	76.78	0.97	0.39	0.27
2.0	1.43	78.87	0.97	0.39	0.27

TABLE 9.4   Function of Gradient α after R. Ehrenberger

α(o)	sinα	sinα$^{0.4}$	sinα$^{-0.26}$	sinα$^{-0.74}$
1	0.017	0.198	2.865	20.000
2	0.035	0.261	2.393	11.976
3	0.052	0.307	2.153	8.873
4	0.070	0.345	1.998	7.174
5	0.087	0.377	1.886	6.084
6	0.105	0.405	1.798	5.318
7	0.122	0.431	1.728	4.747
8	0.139	0.454	1.686	4.303
9	0.156	0.476	1.620	3.946
10	0.174	0.496	1.576	3.653
15	0.259	0.582	1.421	2.719
20	0.342	0.651	1.322	2.212
25	0.423	0.709	1.251	1.891
30	0.500	0.758	1.197	1.670
35	0.574	0.801	1.155	1.509
40	0.643	0.838	1.121	1.387
45	0.707	0.871	1.094	1.292

REFERENCES

1. Abe, I., Odagiri, M. and Ono, K., Studies on the effects of windbreak, *Journal of Agric. Met.* 16, No.1, Tokio, 1960.
2. Alpatyev, A. M., *Vlagooboroty v prirode i ich preobrazovaniya*, Gidrometeoizdat, Leningrad, 1969.
3. Blundell, S. B., Evaporation to leeward of shelterbelt, *Agricult. Meteorology*, 13, 1974, No. 3.
4. Cablík, J., *Construction of Ponds and Utility Water Reservoirs*, SZN, Prague, 1960.
5. Cablík, J. and Jůva, K., *Soil Erosion Control*, SZN, Prague, 1963.
6. Dýrová, E., *Protection and Organization Catchment* (Text-Book), SNTL, Prague, 1974.
7. Ehrenberger, R., *Eine neue Geschwindigkeitsformel für künstliche Gerinne mit starker Neigung und Berechnung der Selbstbelüftung des Wassers*, SWW, H 28/29, 1930.
8. Gabriels, D., Boodt De M. and Minjaw, D., Dune sand stabilization with synthetic soil conditioners: A laboratory experiment, *Soil Science*, Vol,118, No.5, 1974.
9. Hampel, R., Forschungsarbeiten und Versuche auf Gebiete der Wildbach – und Lawinenverbauung, *Allg. Forstzeitung* Nr.23/24, Wein, 1954.
10. Hendrick, R. M. and Mowry, D. T., Effect of synthetic polyelectrolytes on aggregation, aeration and water relationships of soil, *Soil Science*, Vol.73, No.6, 1952.
11. Holý, M., *Contribution to the Solution of the Distance of Infiltration Belts*, ČSAZ No.6, Prague, 1955.
12. Holý, M. *et al.*, *Irrigation Structures*, SNTL, Prague, 1976.

13. Kasprzak, K., *Vegetation Structures and their Use in Soil Conservation*, Prague, 1959.
14. Kolář, V. *et al.*, *Hydraulics - Technical Guide*, SNTL, Prague, 1966.
15. Kondralinov, A. R. and Voroncov, P. A., *Vliyanie lesnykh polos na veter i turbulentniy obmen v atmosfere*, Trudy Gidromet. Inst., No.26, Kyjev, 1961.
16. Kondrashov, S. K., *Oroshayemoye zemledeliye*, Moscow, 1948.
17. Kozmenko, A. S., *Borba s eroziyey pochv*, Moscow, 1954.
18. Kutílek, M., Effects of City Waste Waters on Soil Structure - Doctorship Thesis, ČVUT, Prague, 1956.
19. Mareš, K., *River Training Project Design* - Text-Book, SNTL, Prague, 1974.
20. Mircchulava, C. E., *Inzhenerniye metody rascheta i prognoza vodnoy erozii*, Moscow, 1970.
21. Mráček, Z. and Krečmer, V., *The importance of the Forest for Human Society*, SZN, Prague, 1975.
22. Riedl, O. and Zachar, D. *et al.*, *Forest Land Improvement Projects*, SZN, Prague, 1973.
23. Sedlák, V., Optimal Variants of Technical Solutions of Erosion Control for Individual Production Areas, *Final Report on Research Project*, VU XVI-I-0-329/1:6:3, Brno, 1975.
24. Shell Solfix tegen Zandverstenivingen, Landbouwmechanisatie, No.10, 1973.
25. Strele, G., *Grundriss der Wildbachverbauung*, Wien, 1934.
26. Sus, N. I., *Eroziya pochvy i borba s nyey*, Moscow, 1949.
27. Urban, V., Synthetic Substances in Erosion Control of Slopes, In: *Sborník "Nové melioračni hmoty pro zúrodňováří máloplodných půd v zemědělství a lesnictví"*, Prague, ÚVTI, 1972.
28. Valentini, C. D., *Del modo di determinare il profilo compensazione e sua importanza nelle sistemazioni idrauliche*, Milan, 1895.
29. Velikanov, M. A., *Gidrologiya sushi*, Leningrad, 1948.
30. Zachar, D., *Soil Erosion*, SAV, Bratislava, 1970.
31. Ziemnicki, S., *Melioracje przeciwerozyjne*, Warszawa, 1968.

10. Economics of Erosion Control

It will be extremely difficult to justify erosion control by traditional approaches. This is because erosion processes cause damage to many branches of the national economy and much of this damage, namely its social consequences, is difficult to express in numerical values. The evaluation of social losses resulting from damage caused to the basic natural resources is lacking. Erosion control should therefore be preceded by an economic evaluation of the enumerable losses caused by erosion. Data obtained from the evaluation should be included in the final decision-making process together with the effects which we are as yet unable to express in numerical values. An analysis of these effects will make the final project effective.

The evaluation of the economic effectiveness of erosion control measures with regard to agricultural production should not only consider the current interests of agricultural enterprises but should also take into account the interests of society manifested in the effort to conserve the soil for future generations, to protect the environment, etc. Environmental control and soil conservation will enforce the introduction of erosion control measures in areas where traditional methods have shown them to be economically inefficient and have been an obstacle to their implementation

The effectiveness of erosion control measures is significantly related to their effect on water resources. Erosion processes which cause water pollution and affect large areas significantly affect water quality. This is especially obvious when new large-scale production and chemicalization is being introduced in agriculture.

In order to meet the demands placed by society on water quantity and quality, water must be acquired, accumulated, transported and preserved or treated. Erosion control should therefore be evaluated with regard to the effect of its cost on water retention storage, transportation, etc., i.e., a comparison should be made of the costs of reducing or preventing the pollution or silting of water courses and reservoirs by erosion control and by other available treatments.

The decision-making process which will lead to the protection of water resources should be based on a detailed economic analysis. This is extremely difficult to do because such an analysis must involve detailed physical and technological data and the economist will have to cooperate closely with water management specialists, agronomists, biologists, chemists, hydrologists, geologists, etc. The decision-making process must definitely be a multidisciplinary venture. The economic analysis of measures for the protection and treatment of water resources will be aimed at meeting the requirements of the individual branches of the national economy for an adequate quantity and quality of water while considering the possibilities of obtaining the water at the lowest cost and still fully satisfying the needs of society.

The differentiated approach to water quality demands is extremely important. Potable water will have a different quality to water for navigation, water for energy production, irrigation, the food industry, cooling water, etc., and techniques for obtaining the required quality will therefore also differ as will the costs. The differentiated approach to these requirements in the catchment area will require a systems analysis to be conducted, not only with the aim of determining the technical parameters of the water resources with regard to their uses, but mainly with the aim of determining the final properties of the water for a certain purpose with regard to the available means for obtaining such properties.

In simplified terms the demands on water quality may be classified into three grades:

Grade one does not put demands on improving water quality, water in this case is usually only a means of transporting wastes of different sorts. Grade two requires the preservation of the given water quality which means that water pollution control measures should be introduced. Should such water be polluted by agricultural production erosion control measures must be implemented strictly. Grade three means upgrading the quality of water which will be used for supplying the population with potable water. This requires the introduction of exacting water treatment technologies and the implementation of strict water pollution control measures.

The requirements on water quality stated in grades two and three will be met using a variety of methods which should be assessed with regard to the desired objective and to costs. That method should be chosen which will prove most feasible with regard to both objective and costs. The silting of a water reservoir may, for instance, be prevented by erosion control measures in the catchment, technical structures in the water course (settling basins), diversion of the sediment-bearing water course, a combination of these techniques or by erosion control measures in the catchment, water treatment at take-off from the reservoir or the combination of both techniques.

The effects of erosion control measures for agricultural production should always be considered.

The evaluation of the protection of water resources with regard to the needs of society should be made considering the economic aspects as well as factors which we are as yet unable to express in numerical values, namely a healthy environment, i.e., recreation in unpolluted water and air, the improvement of the aesthetic appearance of the landscape, soil conservation for future generations, etc. Soil as a natural resource has been evaluated in detail in the European Charter on Soil drawn up by the UN.

All measures taken for the protection of water resources, including erosion control, should be carried out with regard to all impacts which they may have on society no matter whether or not it is currently in our power to express them in present economic terms.

REFERENCES

1. Holý, M., Říha, J. and Sládek, J., *Society and the Environment*, Prague, 1975.
2. Holý, M., *Erosion Control*, SNTL/ALFA, Prague, 1978.
3. Wilbrich, T. L. and Smith, G. E., *Agricultural Practices and Water Quality*, Iowa, USA, 1970.